T0133742

The Craft of Model-Based Testing

The Craft of Model-Based Testing

Paul C. Jorgensen

CRC Press
Taylor & Francis Group
Boca Raton London New York

CRC Press is an imprint of the
Taylor & Francis Group, an **informa** business

AN AUERBACH BOOK

CRC Press
Taylor & Francis Group
6000 Broken Sound Parkway NW, Suite 300
Boca Raton, FL 33487-2742

International Standard Book Number-13: 978-1-4987-1228-6 (Hardback)

Library of Congress Cataloging-in-Publication Data

Names: Jorgensen, Paul, author.
Title: The craft of model-based testing / author, Paul Jorgensen.
Description: Boca Raton : Taylor & Francis, a CRC title, part of the Taylor &
Francis imprint, a member of the Taylor & Francis Group, the academic
division of T&F Informa, plc, [2017] | Includes bibliographical references.
Identifiers: LCCN 2016047277 | ISBN 9781498712286 (hb : acid-free paper) |
ISBN 9781315204970 (e)
Subjects: LCSH: Computer simulation. | Testing.
Classification: LCC QA76.9.C65 J665 2017 | DDC 003/.3--dc23
LC record available at https://lccn.loc.gov/2016047277

Visit the Taylor & Francis Web site at
http://www.taylorandfrancis.com

and the CRC Press Web site at
http://www.crcpress.com

To my wife, Carol, our daughters, Kirsten and Katia,
and the grandchildren from A to Z:
Abbie, Gillian, Hope, Max, Nora, Vivian, and Zoe.

Contents

PART 2 THE PRACTICE OF MODEL-BASED TESTING

Preface

First, a disclaimer. I use the words "craftsman" and "craftsmanship" in the gender neutral sense. No offense is meant, and I hope none is taken. I believe model-based testing (MBT) can be, and should be, a craft rather than an art. Craftsmanship entails three essential parts: understanding the medium, ability to choose appropriate tools, and the experience to use them well. The relation between tools and craft is interesting—a craftsman can do acceptable work with poor tools, but a novice cannot do good work with excellent tools. This is absolutely true for MBT as a craft.

Other than software testing, my preferred craft is woodworking. As a craft, a woodworker needs to know the medium, which, in this case, is wood. Different woods have different properties, and knowing these lets a woodworker make appropriate choices. Maple is extremely hard and requires very sharp tools. Pine is very soft, and forgiving. My preferred wood is Cherry—it is not as hard as maple or oak, but it has a fine grain and it "works" well. The tool part is obvious—take hand saws; for example, a craftsman would have crosscut and ripping saws, a back saw, a miter box, a scroll saw, and maybe specialized Japanese saws that cut on the pull stroke. Each of these is good for a special purpose, and there is no one saw that fits all of these purposes. But just having tools is not enough. A would be craftsman must know how to use the tools at his or her disposal. This is where experience comes in. Maybe the best historical example of craft is the guild system, with apprentices, journeymen, and master craftsmen. The whole point of that progression was that, to be recognized as a dependable craftsman, an individual had to go through a long, supervised learning process.

How does all this fit with model-based testing (MBT)? And what constitutes a craftsman-like approach to MBT? The medium is the software (or system) to be tested. In a simple distinction, the software may be either transformational or reactive. Just this difference has an implication for choosing the appropriate MBT tool.

The "tools" part of MBT as a craft consists of the models used to describe the software, covered in Part 1 and the products, commercial or open sourced, that generate and possibly execute test cases derived from the model, covered in Part 2. In Part 1 ("Theory of Models for Model-Based Testing"), after an introductory overview, Chapters 2 through 10 present nine models of varying complexity and expressive power. Some of these are well known—flowcharts and decision tables.

Finite state machines receive extended treatment because they are the model most extensively supported by both commercial and open sourced MBT products. Part 2 (The Practice of Model-Based Testing) presents six commercial MBT products and a final chapter that sketches six open source MBT tools.

Communicating the experience was the most challenging part of writing this book. There are two continuing examples—the Insurance Premium Problem is a transformational application and the Garage Door Controller is a reactive (event-driven) example. These two problems are modeled in an educational way in Chapters 2 through 10. They were also given to the six commercial tool vendors to show how their products support the two continuing examples. The whole MBT community agrees that the success of MBT is largely determined by how well the system to be tested is modeled, hence the importance of Chapters 2 through 10.

My father was a tool and die maker, his father and grandfather were Danish cabinet makers, my other grandfather was a painter, and my wife is an excellent cook. These family members all approach their talents as crafts, and they always had pride in their work. I believe that sense of pride helps elevate ordinary work to a craft. My goal for readers of this book is that they (you) can use the material here to become an MBT craftsperson.

Paul C. Jorgensen
Rockford, MI

Acknowledgments

I am very thankful for the contributions that my colleagues and graduate students have made to this book. Among MBT vendors, thanks to Anne Kramer of sepp. med GmbH, Bruno Legeard of Smartesting, Andrus Lehtmets, the CEO of Elvior, LLC, Yaxiong Lin, the CEO of TestOptimal LLC, Prof. Jan Peleska of the University of Bremen, Germany, and RTtester, and Stephan Schulz, the chief technical officer of Conformiq.

Thanks also to my teams of graduate students at Grand Valley State University, Allendale, Michigan: Mohamed Azuz, Khalid Alhamdan, Khalid Almoqhim, Sekhar Cherukuri, James Cornett, Lisa Dohn, Ron Foreman, Roland Heusser, Ryan Huebner, Abinaya Muralidharan, Frederic Paladin, Jacob Pataniczak, Kyle Prins, Evgeny Ryzhkov, Saheel Sehgal, Mike Steimel, Komal Sorathiya, and Chris Taylor.

At Grand Valley State University, the director of the School of Computing and Information Systems, Dr. Paul Leidig, the dean of the Padnos College of Engineering and Computing, Dr. Paul Plotkowski, and the university provost, Dr. Gayle Davis, all approved my sabbatical leave, which gave me the time to write this book.

Finally, I express deep gratitude to my wife, Carol, who has been more than patient with me for the past many months.

About the Author

Paul C. Jorgensen, PhD, spent 20 years of his first career in all phases of software development for telephone switching systems. He began his university career in 1986 teaching graduate courses in software engineering at Arizona State University, Tempe, Arizona, and since 1988 at Grand Valley State University, Allendale, Michigan, where he is a full professor. His consulting business, Software Paradigms, hibernates during the academic year, and emerges for a short time in the warmer months. He has served on major COnference on DAta SYstems Languages (CODASYL), Association for Computing Machinery (ACM), and Institute of Electrical and Electronics Engineers (IEEE) standards committees, and in 2012, his university recognized his lifetime accomplishments with its "Distinguished Contribution to a Discipline Award."

In addition to the fourth edition of his software testing book, *Software Testing: A Craftsman's Approach*, he is also the author of *Modeling Software Behavior: A Craftsman's Approach*. He is a coauthor of *Mathematics for Data Processing* (McGraw-Hill, 1970) and *Structured Methods—Merging Models, Techniques, and CASE* (McGraw-Hill, 1993). More recently, Dr. Jorgensen has been involved with the International Software Testing Certification Board (ISTQB) where he is a coauthor of the advanced level syllabi and served as the vice-chair of the ISTQB Glossary Working Group. He was a reviewer for the ISTQB Model-Based Testing syllabus.

Living and working in Italy for three years made him a confirmed "Italophile." He, his wife Carol, and daughters Kirsten and Katia have visited friends there several times. Paul and Carol have volunteered at the Porcupine School on the Pine Ridge Reservation in South Dakota every summer since 2000. His university email address is jorgensp@gvsu.edu, and when he becomes a professor emeritus in the summer of 2017, he can also be reached at pauljorgensen42@gmail.com.

THEORY OF MODELS FOR MODEL-BASED TESTING

1

Chapter 1

Overview of Model-Based Testing

All testing, software, hardware, or in daily life, consists of checking the response that comes from a stimulus. Indeed, one of the early requirements methods focused on stimulus–response pairs. In model-based testing (MBT), we consider models that express, to some extent, the stimuli and responses of a system we wish to test. Some clarifying vocabulary will help our discussion.

There are three generally accepted levels of testing: unit, integration, and system. Each of these has distinct goals and methods. Unit testing occurs at the class or procedure level, integration testing considers sets of interacting units, and system testing occurs at the port boundary of a system under test (SUT). There are test coverage metrics that apply to each of these levels. (For a more complete discussion, see [Jorgensen 2013].) At any of these levels, a test case consists of some identification name and identifier, preconditions for test execution, a sequence (possibly interleaved) of inputs and expected outputs, a place to record observed outputs, postconditions, and a pass/fail judgment.

1.1 Initial Terminology

Definition: A *system under test*, usually abbreviated as SUT, is a system being tested.

A SUT can be a full system of software-controlled hardware, a system of hardware alone, or software alone, or even a system of SUTs. A SUT can also be a single software unit, or a collection of units.

Definition: The *port boundary* of a system under test is the set of all points at which input stimuli and output responses occur.

Every system, hardware, software, firmware, or some combination of these, has a port boundary. Identifying the "true" port boundary of a SUT is essential to the MBT process. Why "true"? It is easy to confuse user-caused physical events with their electronic recognition (stimuli). In a web-based application, the user interface is likely the location of both system-level inputs and outputs. In an automobile Windshield Wiper Controller, the port boundary typically includes a lever and a dial to determine wiper speeds and a motor that drives the wiper blades. Many examples in this book use a Garage Door Controller (more completely defined later). The port boundary of this system includes devices that send a control signal, safety devices, end-of-track sensors, and a driving motor. The port boundary of a unit is the mechanism by which the unit is activated (a message in object-oriented software, or a procedure call in traditional software).

Definition: A port input event is a stimulus that occurs at the port boundary of a given SUT; similarly, a port output event is a response that occurs at the SUT port boundary.

There is a curious potential point of confusion between developers and testers. Consider port input events. A tester thinks in terms of generating, or causing, inputs to a SUT, whereas a developer thinks more in terms of sensing and acting on them. This dichotomy obviously applies to output events, which are observed or sensed by a tester, and caused or generated by a developer. Part of this confusion is the result of design and development models that were created by the software development community. This becomes a problem for MBT if the underlying model uses the developer viewpoint, but a test case uses the tester viewpoint.

1.2 Events

The following terms are roughly synonymous, but they need clarification: port input event, stimulus, and input; symmetrically, we have the following synonyms: port output event, response, and output. Events occur in "layers", maybe "sequences" is a better term. For now, consider the Garage Door Controller example (see Section 1.8.2 for a full description), specifically the light beam sensor safety device. When the door is closing, if anything interrupts the light beam (near the floor), the motor immediately stops and reverses direction to open the door. The "event sequence" begins with some physical event, perhaps an animal crossing the path of the beam while the door is closing. When the light beam sensor detects the interruption, it sends a signal to the controller; this is a port input event and is a true electronic signal. The software internal to the controller would consider this to be a logical event.

Port input events may occur in different logical contexts. The physical event of a cat crossing the light beam can occur in several contexts: when the door is open, when it is opening, or when it is closing. The logical event only occurs when the door is closing. Frequently, event contexts are represented as states in some finite state machine (FSM). As we examine various models for the way they support MBT, the ability to represent, and recognize, context sensitive input events will be important. Also this forces attention to the port input devices themselves. Suppose, for example, that a tester wanted to test the light beam sensor, particularly its failure modes. The common device failures are Stuck-at-1 (SA-1) and Stuck-at-0 (SA-0). With a SA-1 failure, the light beam sensor will ALWAYS send a signal, regardless of a physical input event that may or may not occur. Note that it will be impossible to close the garage door with this fault. (See use case EECU-SA-1 in the table given below.) The SA-0 fault is more insidious—the door will not reverse after the physical interruption. I am sure the lawyers will get very upset about SA-0 faults on a safety device. It will be modeled in Chapter 8.

Use Case Name	Light Beam Sensor Stuck-at-1.
Use Case ID	EEUC-SA-1.
Description	A customer attempts to close an open door with a control device signal. The Light Beam Sensor has a SA-1 fault.
Preconditions	1. The garage door is open.
	2. The Light Beam Sensor has a SA-1 fault.
Event Sequence	
Input events	Output events
1. Control device signal.	2. Start motor down.
3. Light beam SA-1 fault.	4. Stop and reverse motor.
5. End of up track reached.	6. Stop motor.
Postconditions	1. The garage door is open.
	2. The Light Beam Sensor has a SA-1 fault.

Stuck-at faults, and indeed other failure modes, are difficult to anticipate. They may or may not appear in the requirement specification. Even if they do, they are hard to model in many MBT models. We will revisit this in great detail in

Chapter 8. Is it likely that a customer would offer use cases such as EEUC-SA-1? Possibly, based on past experience, but this would be a challenge in an agile development.

1.3 Test Cases

There are two fundamental forms of a test case—abstract and real (some parts of the MBT community refer to the latter as "concrete test cases"). An abstract test case can usually be derived from a formal model; what makes it "abstract" is that the inputs are usually expressed as variables. A real (concrete) test case contains actual values of input variables and values of expected output values. Both forms should include pre- and post-conditions.

1.4 An Architecture for Test Case Execution

Figure 1.1 sketches a generalized architecture for automated test case execution. It is based on a system my team developed for regression testing of telephone switching systems in the early 1980s. The computer houses the overall test case processor, which controls and observes test case execution. Test cases are expressed in a simple language that can be executed interpretively. The language consists of CAUSE and VERIFY statements that refer to the port boundary of the SUT. A CAUSE statement typically has parameters that refer to port input events and the devices where they occur (these may have additional parameters). Similarly, VERIFY statements refer to expected port output events. In a telephone SUT, we might have the following two statements in a test case:

CAUSE InputDigit(9) on Line12
VERIFY DigitEcho(9) on Line12

In these statements, InputDigit refers to a parameterized port input event that occurs on the device named Line12 and a port output event that occurs on the

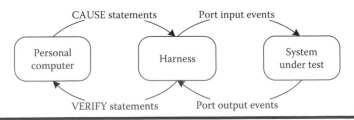

Figure 1.1   A generic test execution architecture.

same device. The key to making this work is to develop a "harness" that connects the test case processor to the SUT. The harness essentially performs logical-to-physical translations of port input events in a CAUSE statement, and physical-to-logical translations of port output events in a VERIFY statement.

All of this is directed at system level testing. Given the popularity of unit level automatic testing support programs such as the nUnit family, the CAUSE and VERIFY statements are replaced by ASSERT statements that contain both unit level inputs and outputs, thereby replacing the harness needed for system testing. In this book, we ignore integration testing, as there is little MBT tool support for this level. This chapter ends with examples of MBT at both the unit and the system levels. For now, it is important to understand that, for MBT to be successful, the underlying models must refer to both stimuli and responses, whether at the unit or the system level.

1.5 Models for MBT

Software (and system) design models are of two general types—structural or behavioral. In the Unified Modeling Language (UML), the *de facto* standard, structural models focus on classes, their attributes, methods, and connections among classes (inheritance, aggregation, and association). There are two main behavioral models—Statecharts and activity (or sequence) diagrams. Part 1 of this book presents nine behavioral models: Flowcharts, Decision Tables, Finite State Machines, Petri nets, Event-Driven Petri Nets, Statecharts, Swim Lane Event-Driven Petri Nets, UML (use cases and activity charts), and the Business Process Modeling and Notation (BPMN). Each of these, except Swim Lane Event-Driven Petri Nets [Jorgensen 2015] and BPMN, is explained and illustrated fully in [Jorgensen 2008]. The focus here is the extent to which these models support MBT. One inescapable limitation of MBT is that the derived test cases are only as good as the information in the model from which they are derived. Thus a recurring emphasis in Part 1 is the expressive power as well as the limitations of the various models.

1.6 The ISTQB MBT Extension

This book is intended to coordinate with the MBT extension to the ISTQB Foundation Level Syllabus, which was released in October 2015. The ISTQB (International Software Testing Qualification Board) is a nonprofit organization that, in 2013, had certified more than 336,000 software testers in 100 countries [http://www.istqb.org/]. Part 2 of this book consists of six vendor-supplied chapters describing their commercial products, and the results of their products on the two continuing examples (defined in Section 1.8).

1.7 Forms of Model-Based Testing

There are three commonly identified forms of model-based testing: manual, semi-automated, and fully automated [Utting 2010]. In manual MBT, a model of the SUT is developed and analyzed to identify test cases. For example, if the SUT is modeled by a finite state machine, paths from initial states to final states can be visually identified and converted into test cases. The next step is to apply some selection criterion to choose which test cases should be executed. The selection criterion is most likely some coverage metric. These will then need to be "concretized" (the popular term in MBT circles) to replace abstract terms, such as Personal ID number with a real value, such as "1234." The final step is to execute the concretized test cases on the SUT. As a side note, Craig Larman [Larman 2001] identifies four levels of Use Cases (more on this in Chapter 9). The third level, Expanded Essential Use Cases, contains abstract variable names; Larman's fourth level, Real Use Cases, replaces the abstract terms with actual values that can be tested. This is exactly the sense of the concretization process. The early use of tools distinguishes manual from semiautomated MBT. Tools are usually some engine that can execute a suitable model and generate abstract test cases. The next step could be either manual or automated: selecting a set of test cases from among the generated set. In a finite state machine example, possible selection criteria can be those test cases that

1. Cover all states.
2. Cover all transitions.
3. Cover all paths.

The selection process can be automated. In fully automated MBT, the steps of semiautomated MBT are followed by automated test case execution. (More on this in Part 2.)

1.8 Continuing Examples

1.8.1 A Unit Level Problem: Insurance Premium Calculation

Premiums on an automobile insurance policy are computed by cost considerations that are applied to a base rate. The inputs to the calculation are as follows:

1. The base rate ($600).
2. The age of the policy holder (16 <= age < 25; 25 <= age < 65; 65 <= age < 90).
3. People less than age 16 or more than 90 cannot be insured.

4. The number of "at fault" claims in the past five years (0, 1–3, and 3–10).
5. Drivers with more than 10 at fault claims in the past five years cannot be insured.
6. The reduction for being a goodStudent ($50).
7. The reduction for being a nonDrinker ($75).

The calculation values are shown in Tables 1.1 through 1.3.

1.8.2 A System Level Problem: The Garage Door Controller

A system to open a garage door is composed of several components: a drive motor, the garage door wheel tracks with sensors at the open and closed positions, and a control device. In addition, there are two safety features: a laser beam near the

Table 1.1   Premium Multiplication Values for Age Ranges

Age Ranges	ageMultiplier
16 <= age < 25	x = 1.5
25 <= age < 65	x = 1.0
65 <= age < 90	x = 1.2

Table 1.2   Premium Penalty Values at Fault Claims

"At Fault" Claims in the Past Five Years	claimsPenalty
0	$0
1 to 3	$100
4 to 10	$300

Table 1.3   Decision Table for goodStudent and nonDrinker Reductions

c1. goodStudent	T	T	F	F
c2. nonDrinker	T	F	T	F
a1. Apply $50 reduction	x	x	—	—
a2. Apply $75 reduction	x	—	x	—
a3. Do nothing	—	—	—	—

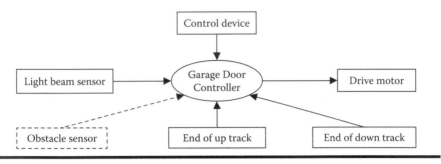

Figure 1.2   SysML diagram of the Garage Door Controller.

floor and an obstacle sensor. These latter two devices operate only when the garage door is closing. While the door is closing, if either the light beam is interrupted (possibly by a pet) or if the door encounters an obstacle, the door immediately stops, and then reverses its direction. To reduce the size of models in subsequent chapters, only the light beam sensor is considered. The corresponding analyses for the obstacle sensor are almost identical. When the door is in motion, either closing or opening, and a signal from the control device occurs, the door stops. A subsequent control signal starts the door in the same direction as when it was stopped. Finally, there are sensors that detect when the door has moved to one of the extreme positions, either fully open or fully closed. When either of these occurs, the door stops. Figure 1.2 is a SysML context diagram of the Garage Door Controller.

In most garage door systems, there are several control devices: a digital keyboard mounted outside the door, a separately powered button inside the garage, and possibly several in-car signaling devices. For simplicity, we collapse these redundant signal sources into one device. Similarly, as the two safety devices generate the same response, we will drop consideration of the obstacle sensor and just consider the light beam device.

1.8.3 Additional Examples

There are several other examples that are used to illustrate and compare modeling theory and techniques: these are deliberately chosen to illustrate model-specific features. Table 1.4 describes the utility of various modeling choices on the examples in Table 1.5.

Table 1.4   Model Choices for the Examples in Table 1.5

	Flowchart	Decision Tables	Finite State Machines	(Ordinary) Petri nets	Event-Driven Petri Nets	Statecharts		
Good Choice	WCF	ND	EVM	EVM	EVM	EVM	WCF	Wind Chill Formula
	IP	WW	RRX	RRX	RRX	RRX	IP	Insurance Premium
				WW	WW	WW	ND	NextDate
				PCP	PCP	PCP	EVM	Espresso Vending Machine
			GDC	GDC	GDC	GDC	RRX	RR Crossing Gate controller
Can Work, But...	ND	IP	WW				WW	Windshield Wiper
	EVM	EVM					GDC	Garage Door Controller
	RRX	RRX					PCP	Producer–Consumer Problem
	WW	GDC						
Poor Choice	PCP	WCF	WCF	WCF	WCF	WCF		
	GDC	PCP	IP	IP	IP	IP		
			ND	ND	ND	ND		
			PCP					

Table 1.5   Chapter-Specific Supplemental Examples Used in Part 1 Chapters

Chapter	Model	Chapter-Specific Supplemental Examples
2	Flowcharts	Espresso Vending Machine (EVM), NextDate (ND), Windchill Formula (WCF)
3	Decision tables	NextDate (ND)
4	Finite state machines	Railroad Crossing Gate Controller (RRX), Windshield Wiper Controller (WW)
5	Petri nets	Producer–consumer problem (PCP)
6	Event-Driven Petri Nets	Windshield Wiper Controller (WW)
7	Statecharts	Windshield Wiper Controller (WW)
8	Swim Lane EDPNs	
9	Unified Modeling Language	

1.9 The MBT Surveys

In its manual form, model-based testing has been in use since the 1980s. The advent of open-source MBT tools (primarily in the academic community) expanded the interest in MBT. More recently, the availability of commercial MBT products has brought the technology into the industrial practice. Some of the motivating factors for adopting MBT are highlighted in two surveys conducted by Robert V. Binder. The initial survey of MBT users was in 2011, and a follow-up survey was made in 2014. As this book goes to press, the 2016 survey is in progress. These surveys summarize the hopes and concerns of the early MBT adopters.

Binder highlights the following observations of the 2011 survey [Binder 2012]:

- "MBT usage spans a wide range of application stacks, software processes, application domains, and development organizations.
- MBT is accessible and practical: half of the respondents report becoming minimally proficient with their MBT tool with 80 or fewer hours of training or coaching; 80% with 100 hours or less.
- On average, respondents report MBT reduced escaped bugs by 59%.
- On average, respondents report MBT reduced testing costs by 17%.
- On average, respondents report MBT reduced testing duration by 25%."

The survey was repeated in 2014, this time two other MBT practitioners, Anne Kramer (see Chapter 18) and Bruno Legeard (see Chapter 14) joined the effort [Binder 2014]. There were exactly 100 responses. The referenced report highlights the following points (quoted directly or paraphrased):

Testing levels

- 77.4% used MBT for system testing.
- 49.5% used MBT for integration testing.
- 40.9% used MBT for acceptance testing.
- 31.2% used MBT for component testing.

Generated artifacts

- 84.2% automated test scripts.
- 56.6% manual test cases.
- 39.5% test data.
- 28.9% other documents.

Biggest benefits

- Test coverage.
- Mastering complexity.

- Automatic test case generation.
- Reuse of models and model elements.

Biggest limitations

- Tool support.
- Skill availability for MBT.
- Resistance to change.

General observations

- 96% used MBT for functional testing.
- 81% used graphical models.
- 59% modeled behavioral aspects.
- Approximately 80 hours needed to become a proficient user.
- 72% of participants were very likely to continue using MBT.

User expectations for MBT

- 73.4% more efficient test design.
- 86.2% more effective test cases.
- 73.4% manage complexity of system testing.
- 44.7% improve communication.
- 59.6% start test design earlier.

Overall effectiveness of MBT

- 23.6% extremely effective.
- 40.3% moderately effective.
- 23.6% slightly effective.
- 5.6% no effect.
- 1.4% slightly ineffective.
- 2.8% moderately ineffective.
- 2.8% extremely ineffective.

References

[Binder 2012]
Binder, Robert V., Real Users of Model-Based Testing, blog, http://robertvbinder.com/real-users-of-model-based-testing/, January 16, 2012.
[Binder 2014]
Binder, Robert V., Anne Kramer, and Bruno Legeard, *2014 Model-Based Testing User Survey: Results*, 2014.

[Jorgensen 2008]
Jorgensen, Paul C., *Modeling Software Behavior—A Craftsman's Approach*. CRC Press, Boca Raton, FL, 2008.
[Jorgensen 2013]
Jorgensen, Paul C., *Software Testing—A Craftsman's Approach*, 4th ed. CRC Press, Boca Raton, FL, 2013.
[Jorgensen 2015]
Jorgensen, Paul C., A Visual Formalism for Interacting Systems. In Petrenko, Schlingloff, Pakulin (Eds.): *Tenth Workshop on Model-Based Testing (MBT-2015), Proceedings MBT 2015*, arXiv:1504.01928, EPTCS 180, 2015, pp. 41–55. DOI:10.4204/EPTCS.180.3.
[Larman 2001]
Larman, Craig, *Applying UML and Patterns: An Introduction to Object-Oriented Analysis and Design*, 2nd ed. Prentice-Hall, Upper Saddle River, NJ, 2001.
[Utting 2010]
Utting, Mark, Pretschner, Alexander, and Legeard, Bruno, A taxonomy of model-based testing approaches. *Software Testing, Verification and Reliability* 2012;22:297–312. Published online in Wiley InterScience (www.interscience.wiley.com). DOI:10.1002/stvr.456.

Chapter 2

Flowcharts

Flowcharts have been in use since the early days of computing; as such they are likely the earliest behavioral model. In the 1960s, vendors often gave away plastic flowchart templates so that programmers could produce neater diagrams. The IBM Corporation even had a standard flowchart template, with varying sizes of the basic symbols. Some old-timers joke that this was the first CASE tool (Computer-Aided Software Engineering).

2.1 Definition and Notation

There are two distinct styles of flowchart notation. The first, shown in Figure 2.1, is minimalist in the sense that it only has symbols for actions, decisions, and two kinds of page connectors. The off-page connector is used for large systems that cannot fit neatly on a single page. The convention for off-page connectors is to use capital letters, A, B, C …, where the first off-page connection would be A, and the point to which the connection is made (on a separate page) is also labeled with an A.

In the early days of flowchart usage, much attention was paid to Input/Output (I/O) devices, and this attention was reflected in an expanded set of flowchart symbols similar to the ones shown in Figure 2.2. (There was even a shape for magnetic tape!)

The "flow" portion of the flowchart notation is shown with arrows emanating from and terminating on separate flowchart symbols. Figure 2.3 is a sample flowchart of an Espresso Vending Machine that dispenses a shot of espresso for one Euro. It will be used to support further discussion of the flowchart technique.

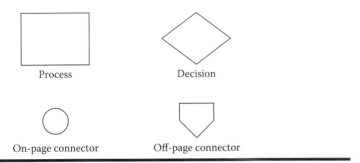

Figure 2.1   Minimalist flowchart symbols.

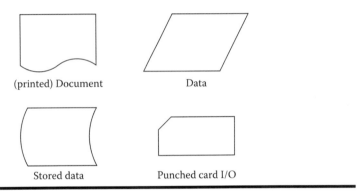

Figure 2.2   I/O device flowchart symbols.

2.2 Technique

The flowchart in Figure 2.3 describes processing on a vending machine that accepts Euro coins and dispenses espresso coffee. If you look carefully at the flowchart, you might notice that arrows generally terminate on the top part of a symbol. This is not mandatory, but it does help comprehension. Decision boxes have exactly one entry point and, because a decision is being made, at least two exit arrows. In the old FORTRAN days, when the language had an "Arithmetic IF" with three outcomes, the sides and bottom vertex of the diamond were used for the three choices (<, =, >). Three decision alternatives are easily shown, as with the Euro coins in Figure 2.3. If more than three alternatives are needed, as in a Case/Switch statement, just use one connecting arrow for each alternative. (Notation is the slave, not the master!) Process boxes, on the other hand, have a maximum of one exit arrow, but they may have several incoming flow arrows. Labels on flow arrows normally refer to the possible outcomes of a decision box: Yes/No, True/False, or values that show outcomes of the decision. Flow arrows emanating from process boxes are not

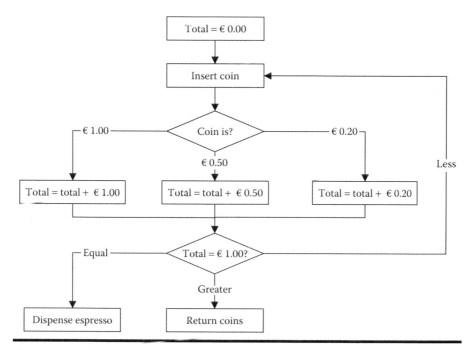

Figure 2.3   Flowchart of the Espresso Vending Machine.

labeled. The intent is that the activity described in the process box is complete, and flow goes to the "next" box. There is no representation of information content flow among boxes. We are asked to imagine that the results of one processing box are "available" (flowcharts have "memory") to subsequent boxes.

There is very little restriction on the textual content of process and decision boxes. There are distinct styles, and these correspond to levels of abstraction. Box text can be quite general, as in Figure 2.3, or it could be quite specific, almost at the programming language construct level. For example, the "Coin is" decision box has alternatives €1.00, €0.50, and €0.20. Alternatively, the decision condition could be expressed with binary conditions for each coin type with emanating choices "Yes" and "No". Whichever choice is made, it is a good practice to remain consistent: mixing levels of abstraction can be confusing. The symbol set for flowcharts explicitly supports the three basic constructs of Structured Programming: sequence, selection, and repetition. The decision boxes in Figure 2.3 are all examples of selection. There is one example of repetition: near the bottom of Figure 2.3, the "less" alternative of the comparison of the variable Total with €1.00 flows back to the insert coin process box. The €1.00 branch, that terminates with the dispense espresso process box is an example of sequence. The Single Entry, Single Exit convention of Structured Programming is not necessarily enforced, but it is followed in Figure 2.3.

The flowchart notation can deal with levels of abstraction by using a form of hierarchy. A high level process box, for example, can be expanded in more detail in a separate flowchart. If this is done, either the lower level flowchart should be named, so that it is clearly a decomposition of the higher level process box, or off-page connectors can be used. (The latter choice is more cumbersome.) Table 2.4, at the end of this chapter, summarizes the control issues that can be represented in flowcharts.

2.3 Examples

2.3.1 The NextDate Function

The NextDate function is popular in testing literature because it is easy to find the expected value of the output portion of a test case. (Given a date, NextDate returns the date of the next day.) Figures 2.4 and 2.5 show how flowcharts represent hierarchy and functional decomposition. The value of the variable lastDay is computed "offline" and then used in comparison with the day variable in the decision box. When a variable gets a value at more than one place, we must ensure that the computed values do not overwrite each other when the flow lines

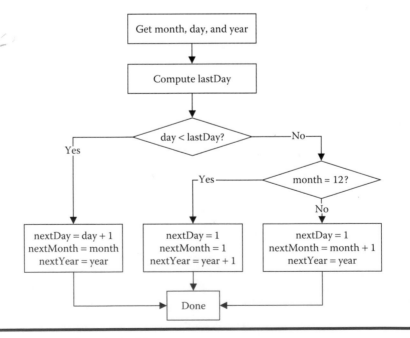

Figure 2.4   NextDate Function.

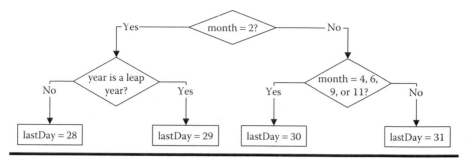

Figure 2.5   Details of lastDay calculation.

are followed. In the NextDate flowchart, there are three assignments of values to nextDay, nextMonth, and nextYear. The decision boxes force these three assignments to be mutually exclusive. Incidentally, these calculations show the apparent use of GoTo statements, a common practice when flowcharts were the dominant behavioral model. The NextDate function is logic-intensive, and this is displayed well in the flowchart.

2.3.2 Wind Chill Factor Table

The famous (in Michigan, and other cold parts of the world) Wind Chill Factor (W) is a function of two variables: wind speed in miles per hour (V) and temperature in Fahrenheit degrees (T). The actual formula is

$$W = 35.74 + 0.6215 * T - 35.75 * (V^{0.16}) + 0.4275 * T * (V^{0.16})$$

where:
W is the apparent temperature on a human face, measured in degrees Fahrenheit
T is the air temperature, measured in degrees Fahrenheit, $-20 <= T <= 50$
V is the wind velocity measured in miles per hour, $3 <= V <= 73$

The flowchart in Figure 2.6 will complete entries in a table similar to Table 2.1. Note the use of nested loops.

(The ranges are arbitrary, but also realistic for Michigan!) The Wind Chill table considers temperatures in the $-20°F <= T <= 50°F$ range at 5 degree increments, and wind speeds in the $3 <= V <= 73$ at 5 m.p.h. increments.

2.3.3 Insurance Premium Calculation Flowchart

Figure 2.7 is a flowchart model of the Insurance Premium Calculation as defined in Chapter 1, Section 1.8.1.

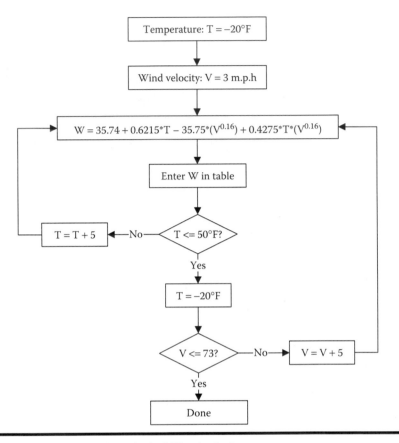

Figure 2.6   Flowchart for Wind Chill calculation.

Table 2.1   Wind Chill as a Function of Wind Velocity and Air Temperature

Temperature/Velocity	−20°F	−15°	...	0°	...	45°	50°
3 m.p.h							
8 m.p.h							
...							
70 m.p.h							
73 m.p.h							

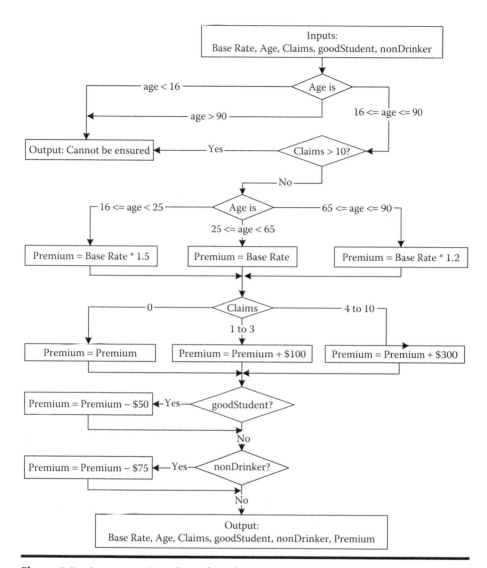

Figure 2.7   Insurance Premium Flowchart.

2.3.4 Garage Door Controller Flowchart

Figure 2.8 is a flowchart model of the problem as defined in Chapter 1, Section 1.8.2.

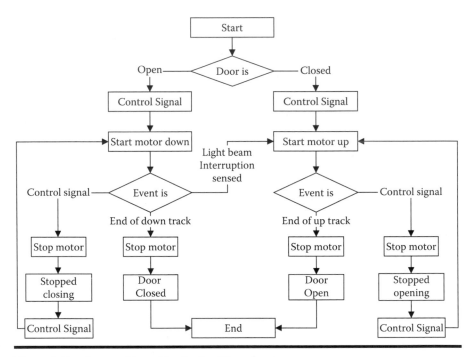

Figure 2.8   Garage Door Controller Flowchart.

2.4 Deriving Test Cases from Flowcharts

Paths in a flowchart lead directly to abstract test cases. As flowcharts can show parallel paths, it is easy to manually derive related abstract test cases. The Insurance Premium Calculation flowchart (Figure 2.7) shows parallel paths for the age and for the claims variables. Parallel paths are understood to be mutually exclusive, although this is usually not explicitly stated. (It can usually be derived from the semantic content of a given flowchart, but this, in turn, mandates manual test case derivation.) In general, there is no possibility of deriving real test cases because a flowchart only indicates what actions need to be taken, it does not actually perform the actions.

Flowcharts are fine for calculation-intensive (data-driven) applications. (They work well for FORTRAN-like programs.) They are less useful for event-driven applications. Looking at the flowchart of the continuing event-driven example, several problems are evident that include the following:

■ Events might be shown as decision outcomes or as actions (process boxes).
■ There are no special symbols for events, so they must be recognized by the semantic content of the flowchart.
■ Insight and domain experience are needed to identify context-dependent input events.

2.4.1 *Insurance Premium Calculation Test Cases*

There are 40 distinct paths through the flowchart in Figure 2.7; 36 of these correspond to Strong Normal Equivalence Class Testing test cases listed in Table 2.2. The age and claims variables are defined by ranges of values, so, on the surface, it would appear that some form of boundary value testing would be appropriate. However, each boundary value in an equivalence class is mapped to the same set of multipliers and additive functions, so the result would be massive redundancy with no additional insight. Figure 2.9 highlights the path through the flowchart that corresponds to test case 1 in Table 2.2.

Table 2.2 shows both the abstract and actual test cases for the Insurance Premium Calculation. The abstract test cases can be derived directly from the flowchart (an MBT tool should be able to do this). The actual values must be derived from the problem definition. With the text as given in Figure 2.7, it would be difficult for an MBT tool to derive the actual values for the test cases. I did most of it by careful use of the Replace function in the spreadsheet I used, so this is not too onerous. As a general rule, spreadsheets are a handy complement to MBT tools.

Here are the abstract and real test cases for test case 1 derived from the path in Figure 2.8:

Insurance Premium Abstract Test Case 1	
Preconditions	Base, Age, Claims, goodStudent, and nonDrinker are known
Inputs	Actions
1. Age in 16 <= age <25	2. Multiply Base by 1.5
3. Claims = 0	4. Multiply Base by 1.0
5. goodStudent is True	6. Subtract $50 from Base
7. nonDrinker is True	8. Subtract $75 from Base
Postconditions	Base contains the premium value

Insurance Premium Real Test Case 1	
Preconditions	Base = $600
Inputs	Actions
1. Age = 20	2. $900 = $600 * 1.5
3. Claims = 0	4. $900 = $900 – $0
5. goodStudent is True	6. $850 = $900 – $50
7. nonDrinker is True	8. $775 = $850 – $75
Postconditions	Premium = $775

Table 2.2 Abstract and Actual Test Cases for the Insurance Premium Problem

Test Case	Input Data				Values (Base rate = $600)			
	Age	Claims	goodStudent	nonDrinker	Age Multiplier	Claims ($)	goodStudent ($)	nonDrinker ($)	Premium ($)
1	20	0	Yes	Yes	1.5	0	−50	−75	775
2	20	0	Yes	No	1.5	0	−50	0	850
3	20	0	No	Yes	1.5	0	0	−75	825
4	20	0	No	No	1.5	0	0	0	900
5	20	2	Yes	Yes	1.5	100	−50	−75	875
6	20	2	Yes	No	1.5	100	−50	0	950
7	20	2	No	Yes	1.5	100	0	−75	925
8	20	2	No	No	1.5	100	0	0	1,000
9	20	6	Yes	Yes	1.5	300	−50	−75	1,075
10	20	6	Yes	No	1.5	300	−50	0	1,150
11	20	6	No	Yes	1.5	300	0	−75	1,125
12	20	6	No	No	1.5	300	0	0	1,200
13	45	0	Yes	Yes	1	0	−50	−75	475
14	45	0	Yes	No	1	0	−50	0	550

(Continued)

Table 2.2 (Continued) Abstract and Actual Test Cases for the Insurance Premium Problem

Test Case	Input Data					Values (Base rate = $600)			
	Age	Claims	goodStudent	nonDrinker	Age Multiplier	Claims ($)	goodStudent ($)	nonDrinker ($)	Premium ($)
15	45	0	No	Yes	1	0	0	−75	525
16	45	0	No	No	1	0	0	0	600
17	45	2	Yes	Yes	1	100	−50	−75	575
18	45	2	Yes	No	1	100	−50	0	650
19	45	2	No	Yes	1	100	0	−75	625
20	45	2	No	No	1	100	0	0	700
21	45	6	Yes	Yes	1	300	−50	−75	775
22	45	6	Yes	No	1	300	−50	0	850
23	45	6	No	Yes	1	300	0	−75	825
24	45	6	No	No	1	300	0	0	900
25	75	0	Yes	Yes	1.2	0	−50	−75	595
26	75	0	Yes	No	1.2	0	−50	0	670
27	75	0	No	Yes	1.2	0	0	−75	645
28	75	0	No	No	1.2	0	0	0	720

(Continued)

Table 2.2 (Continued)   Abstract and Actual Test Cases for the Insurance Premium Problem

Test Case	Input Data				Values (Base rate = $600)			
	Age	Claims	goodStudent	nonDrinker	Age Multiplier	Claims ($)	goodStudent ($)	nonDrinker ($)	Premium ($)
29	75	2	Yes	Yes	1.2	100	−50	−75	695
30	75	2	Yes	No	1.2	100	−50	0	770
31	75	2	No	Yes	1.2	100	0	−75	745
32	75	2	No	No	1.2	100	0	0	820
33	75	6	Yes	Yes	1.2	300	−50	−75	895
34	75	6	Yes	No	1.2	300	−50	0	970
35	75	6	No	Yes	1.2	300	0	−75	945
36	75	6	No	No	1.2	300	0	0	1,020
37	15 or 91	<10	(any)	(any)	(not allowed)				(no premium)
38	20	11	(any)	(any)	(not allowed)	11			(no premium)
39	45	11	(any)	(any)	(not allowed)	11			(no premium)
40	75	11	(any)	(any)	(not allowed)	11			(no premium)

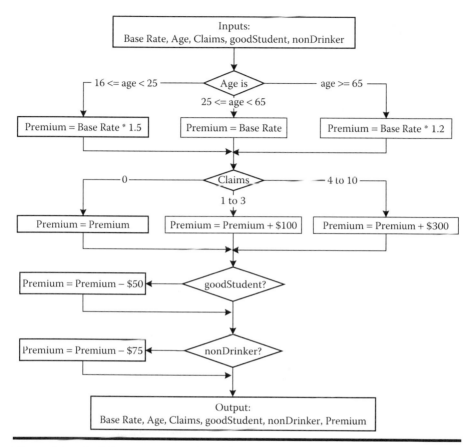

Figure 2.9   Path for Test Case 1 in Table 2.2.

2.4.2 Garage Door Controller Test Cases

There are two loops in the Garage Door flowchart—one when stopping and restarting a closing door, and the other when stopping and restarting an opening door. If we assume that input events and output actions happen instantaneously, a mathematician would say there is a countably infinite set of distinct paths through the flowchart in Figure 2.8. My garage door takes approximately 13 seconds to close (or open), and the stop/restart sequence requires about 1 second, so as a practical matter, there is only a finite number of possible paths in the Garage Door Flowchart. Figure 2.10 highlights one path, and that path is expressed as the following test case. The flowchart symbols in Figure 2.10 are numbered to facilitate path tracing and naming.

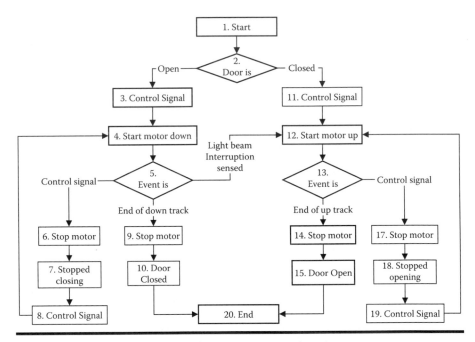

Figure 2.10 One path through the Garage Door Flowchart.

Garage Door Test Case (from Figure 2.10) Flowchart Symbol Sequence: 1, 2, 3, 4, 5, 12, 13, 14, 15, 20	
Preconditions	*Garage Door is Open*
Input Events	*Output actions*
1. Control Signal occurs	2. Start motor down
3. Light Beam Interruption sensed	4. Start motor up
5. End of up track	6. Stop motor
Postconditions	Garage Door is Open

Table 2.3 contains a list of five distinct paths though the Garage Door flowchart. Each of these paths is briefly described (in what might be a user story) and then detailed as a sequence of numbered flowchart symbols. The five paths traverse every flowchart symbol and every edge in the flowchart. There are dozens of other possible paths, obtained from repeated intermediate stops, and also from when the light beam interruption occurs. They also include the control device input event in every context in which it can occur. It would be very difficult for a tool to derive more detailed, use case-like test cases from the flowchart.

Table 2.3   Sample Paths in the Garage Door Flowchart

Sample Garage Door Flowchart Paths		
Path	*Description*	*Flowchart Symbol Sequence*
1	Normal door close	1, 2, 3, 4, 5, 9, 10, 20
2	Normal door close with one intermediate stop	1, 2, 3, 4, 5, 6, 7, 8, 4, 5, 9, 10, 20
3	Normal door open	1, 2, 11, 12, 13, 14, 15, 20
4	Normal door open with one intermediate stop	1, 2, 11, 12, 13, 17, 18, 19, 12, 13, 14, 15, 20
5	Closing door with light beam interruption	1, 2, 3, 4, 5, 12, 13, 14, 15, 20

Table 2.4 contains test case information manually derived from the flowchart in Figure 2.8. The flowchart definition is problematic for the following reasons:

1. Input events appear both as decision outcomes and as actions (in process boxes).
2. Event contexts (states) appear as actions (again, in process boxes) and as decision outcomes.
3. There are two loops that are potentially infinite: the stopping while closing or opening.

There is no support, other than domain experience, to help identify the "next context" that results from an input event/output action pair.

2.5 Advantages and Limitations

Flowcharts have several clear advantages. Any notation that has been used for decades must have something going for it. In the case of flowcharts, it is ease of comprehension. As text in process and decision boxes can be natural language, flowcharts improve understanding between customers and developers. (Even the U.S. Internal Revenue Service uses flowcharts to explain complicated parts of the tax code.) As noted earlier, flowcharts express the basic structured programming constructs. There is also an implicit, if unstated, use of memory. If a variable is given a value in a process and appears sequentially later, the value is assumed to persist. This is prominent in Figure 2.3 in which the variable "total" is defined and redefined in the looping that may occur. Any "well formed" flowchart can be coded in an imperative and structured programming language. As we have seen in our

Table 2.4 Garage Door Test Case Information

Door Context	Input Event	Output Action	Next Context
Open	Control device signal	Start drive motor down	Closing
Closing	End of down track reached	Stop motor	Closed
Closing	Control device signal	Stop motor	Stopped closing
Closing	Light beam interrupted	Reverse motor down to up	Opening
Stopped closing	Control device signal	Start drive motor down	Closing
Closed	Control device signal	Start drive motor up	Opening
Opening	Control device signal	Stop motor	Stopped opening
Opening	End of up track reached	Stop motor	Open
Stopped opening	Control device signal	Start drive motor up	Opening

examples, flowcharts support levels of abstraction; therefore they can scale-up to describe large, complex applications. Another advantage is that they can be used to describe complex computations and algorithms. Finally, some control (or behavior) is represented by the different paths through a flowchart.

There are limitations to flowcharts. The very nature of sequence makes it extremely difficult to describe event-driven systems in which independent events can occur in any order. Second, there is no simple way to describe the context of external devices in which the system being described operates. (The device-oriented I/O symbols might work for this purpose, but there would be many extensions.) There is very little representation of data, except in process and I/O boxes. The text in process and decision boxes can contain variable names, but this is very rudimentary. As data representation is scant, there is even less opportunity to express data structure and relationships among data. There is also potential confusion about describing events. In Figure 2.8, the control device signal sometimes appears as a process box, and other times is an outcome of a decision. Table 2.5 relates flowchart descriptive capabilities to the standard list defined in Chapter 1.

Table 2.5   Representation of Behavioral Issues with Flowcharts

Issue	*Represented?*	*Comment*
Sequence	Yes	Main point of flowcharts
Selection	Yes	Main point of flowcharts
Repetition	Yes	Main point of flowcharts
Enable	No	Must describe as text in a process box
Disable	No	Must describe as text in a process box
Trigger	No	Must describe as text in a process box
Activate	No	Must describe as text in a process box
Suspend	No	Must describe as text in a process box
Resume	No	Must describe as text in a process box
Pause	No	Must describe as text in a process box
Conflict	No	
Priority	No	Must describe as text in a process box
Mutual exclusion	No	Parallel paths after a decision
Concurrent execution	No	Must describe as text in a process box
Deadlock	No	
Context-sensitive input events	No	Must be deduced by careful examination of sequences following a decision
Multiple-context output events	Indirectly	Must be deduced by careful examination of sequences following a decision
Asynchronous events	No	
Event quiescence	No	
Memory?	Yes	By reference to *previous* decisions, inputs, and process boxes
Hierarchy?	Yes	Process boxes and be expanded into more detail as needed

2.6 Lessons Learned from Experience

In the late 1960s, a telephone switching systems development laboratory was required to provide flowchart documentation of all switching system source code sent to the operating companies. At the time, the source code for a telephone office was about 300,000 lines of assembly code. The process in place involved a software designer submitting a hand-drawn flowchart to the drafting department, and *six weeks later*, the nicely drawn, perfectly lettered version was returned to the designer. In the meantime, if the designers made any changes, they could replace the original sketch with a replacement sketch and not lose their place in the drafting department queue.

At about the same time, one of the mathematicians in my supervisory group attended a conference where he saw a program that would draw a discrete component circuit on a CalComp® plotter. We examined the technical material, and decided we could replace the circuit symbols with flowchart symbols. The result was the AELFlow system [Jorgensen and Papendick 1970]. This is where the lessons learned part begins. To get approval for the plotter, we had to show every department how they could benefit from the plotting machine. After weeks of salesmanship, we finally had approval to buy the smallest plotter. Six months later, we had the biggest plotter. The AELFlow system was remarkably effective, both in terms of reduced turn-around time and overall utility of design documentation. (Flowcharts were a much easier subject of a technical inspection than the associated assembly language code.) In many ways, the AELFlow system was a true CASE tool, long before the term "CASE" was in use.

The main lesson is that change is hard to introduce—it takes time, patience, and knowledge of the needs of the organization. It also takes training, although this was minimal for the AELFlow system. The Gartner Hype cycle is pretty accurate, although the durations of the intervals may vary (Figure 2.11). Organizations

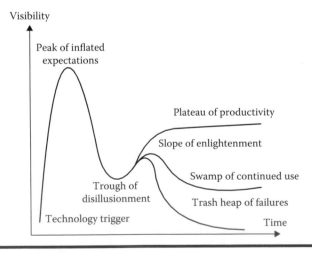

Figure 2.11 Gartner Hype Cycle.

that introduce MBT will experience the hype cycle. In my opinion, the Peak of Inflated Expectations is due to the MBT product sales groups, and the Trough of Disillusionment is due to inadequate training/education on the various models. The Slope of Enlightenment begins with appropriate tool selection and good modeling education, and the Plateau of Productivity can continue with measurement and attention to the marketplace.

Reference

[Jorgensen and Papendick 1970]
Jorgensen, Paul C. and David L. Papendick, AELFLOW—An automated drafting system. *Automatic Electric Technical Journal* 1970;12(4):172–180.

Chapter 3

Decision Tables

Decision tables have been used to represent and analyze complex logical relationships for decades. Their rigorous formulation supports analyses of a crude form of completeness, and absolute forms of redundancy and consistency. In addition, they can support compilation under specific circumstances [CODASYL 1978]. They are ideal for describing situations in which a number of combinations of actions are taken under varying sets of conditions. Decision tables also form a skeleton to guide requirements elicitation between customers and developers. Some of the basic decision table terms are illustrated in Table 3.1.

3.1 Definition and Notation

A decision table has four portions: the part to the left of the bold vertical line is the stub portion; to the right is the entry portion. The part above the bold horizontal line is the condition portion, and below is the action portion. Thus, we can refer to the condition stub, the condition entries, the action stub, and the action entries. A column in the entry portion is a rule. Rules indicate which actions, if any, are taken for the circumstances indicated in the condition portion of the rule. In the decision table in Table 3.1, when conditions c1, c2, and c3 are all true, actions a1 and a2 occur (rule 1). When c1 and c2 are both true and c3 is false, actions a1 and a3 occur (rule 2).

When we have binary conditions (true/false, yes/no, 0/1), the condition portion of a decision table is a truth table (from propositional logic) that has been rotated 90°. This structure guarantees that we consider every possible combination of condition values. Decision tables are deliberately declarative (as opposed to imperative); no particular order is implied by the conditions, selected actions do not occur in any particular order, and rules can be written in any order.

Table 3.1 Parts of a Decision Table

Stub	Rule 1	Rule 2	Rule 3	Rule 4	Rule 5	Rule 6	Rule 7	Rule 8
c1	T	T	T	T	F	F	F	F
c2	T	T	F	F	T	T	F	F
c3	T	F	T	F	T	F	T	F
a1	x	x			x			
a2	x							
a3		x				x	x	x
a4			x	x	x		x	x

This latitude is sometimes handy to express normal cases as the first few rules, and exceptional cases later.

Definition: Decision tables in which all the conditions are binary are called *limited entry decision tables*. We often write LEDT.

A limited entry decision table with n conditions has 2^n distinct rules.

Definition: Decision tables in which all the conditions have a finite number (>2) of alternative values are called *extended entry decision tables*. We often write EEDT.

Definition: Decision tables in which some conditions have a finite number of alternative values and others are strictly binary are called *mixed entry decision tables*. We often write MEDT.

Decision tables presume that all the values necessary to evaluate conditions are available at the onset of table rule execution. Decision table actions may be used to change values of variables, and this may occur in conjunction with a "repeat table" action. There is also a notion of hierarchy: actions may invoke other decision tables.

3.2 Technique

The rigorous structure of decision tables supports some interesting algebraic manipulations.

3.2.1 Decision Table Condensation

When two (or more) rules have identical action entries, there must be a condition that is true in one rule and false in another. Clearly then, that condition has no effect on the actions that are performed in those rules. Table 3.2 condenses rules 3 and 4, and also rules 7 and 8 from Table 3.1.

Table 3.2   Condensing a Decision Table

Stub	Rule 1	Rule 2	Rules 3 and 4	Rule 5	Rule 6	Rules 7 and 8
c1	T	T	T	F	F	F
c2	T	T	F	T	T	F
c3	T	F	—	T	F	—
a1	x	x		x		
a2	x					
a3		x			x	x
a4			x	x		x
Rule count	1	1	2	1	1	2

Definition: If a condition has no effect on the actions performed by two rules, the rule entry for that condition is a "don't care" entry usually written as a long dash (—).

The entry for c3 in rules 3 and 4 is a "don't care" entry. The don't care entry has two major interpretations: the condition is irrelevant, or the condition does not apply. Sometimes people will enter the "n/a" symbol for this latter interpretation. In Table 3.1 the same actions occur in rules 3 and 4; so condition c3 has no effect on the action set. It is replaced with the "don't Care" entry (—); similarly for rules 7 and 8.

A little Boolean algebra explains this more theoretically: as the rules in a well-formed decision table are mutually exclusive, the first simplification is better understood as

> *Rule 3*: $(c1 \wedge (\sim c2) \wedge c3) \rightarrow a4$, and Rule 4: $(c1 \wedge (\sim c2) \wedge (\sim c3)) \rightarrow a4$,
> so their mutual exclusion is
> $((c1 \wedge (\sim c2) \wedge c3) \rightarrow a4)) \oplus ((c1 \wedge (\sim c2) \wedge (\sim c3)) \rightarrow a4)$.
> This reduces to
> $((c1 \wedge (\sim c2)) \wedge (c3 \oplus (\sim c3)) \rightarrow a4$.
> Since $(c3 \oplus (\sim c3))$ is always true we have $(c1 \wedge (\sim c2)) \rightarrow a4$.

3.2.2 Decision Tables with Mutually Exclusive Conditions

When conditions refer to equivalence classes, decision tables have a characteristic appearance. Conditions in the decision table in Table 3.3 refer to part of a calendar problem; they refer to the mutually exclusive equivalence classes for the month variable. As they are mutually exclusive, we cannot ever have a rule in which two entries are true. Unfortunately, the accepted symbol for this is still the don't care entry (—). In such cases, it really means "must be false". Some decision table aficionados use the notation F! to emphasize this point.

Table 3.3 Decision Table with Mutually Exclusive Rules and Rule Counts

Conditions	Rule 1	Rule 2	Rule 3
c1. 30-day month?	T	—	—
c2. 31-day month?	—	T	—
c3. February?	—	—	T
a1			
Rule count	4	4	4

Use of don't care entries has a subtle effect on the way in which complete decision tables are recognized. For limited entry decision tables, if n conditions exist, there must be 2^n distinct rules. When don't care entries really indicate that the condition is irrelevant, we can develop a rule count as follows: rules in which no don't care entries occur count as one rule, and each don't care entry in a rule doubles the count of that rule. The rule counts for the condensed rules in Table 3.2 are shown in the bottom row, and their sum is 2^3. The rule counts for the decision table in Table 3.3 are shown in the bottom row of the table. Notice that the sum of the rule counts is 12 (as it should be).

If we applied this simplistic algorithm to the decision table in Table 3.3, we get the rule counts shown in Table 3.4. We should have only eight rules, so we clearly have a problem. To see where the problem lies, we expand each of the three rules, replacing the "—" entries with the T and F possibilities, as shown in Table 3.5.

Notice that we have three rules in which all entries are T: rules 1, 5, and 9. We also have six rules in which two conditions are true: rules 2, 3, 6, 7, 10, and 11. If we delete these impossible rules, we end up with three rules. The result of this process is shown in Table 3.6 in which we eliminate the impossible rules, and use the "F!" (must be false) notation to emphasize the mutual exclusion.

Table 3.4 Expanded Decision Table for Table 3.3 with Rule Counts

Conditions	Rule 1	Rule 2	Rule 3	Rule 4	Rule 5	Rule 6	Rule 7	Rule 8	Rule 9	Rule 10	Rule 11	Rule 12
c1. 30-day?	T	T	T	T	T	T	F	F	T	T	F	F
c2. 31-day?	T	T	F	F	T	T	T	T	T	F	T	F
c3. February?	T	F	T	F	T	F	T	F	T	T	T	T
a1												
Rule count	1	1	1	1	1	1	1	1	1	1	1	1

Table 3.5   Expanded Decision Table for Table 3.3 with Impossible Rules

Conditions	Rule 1	Rule 2	Rule 3	Rule 4	Rule 5	Rule 6	Rule 7	Rule 8	Rule 9	Rule 10	Rule 11	Rule 12
c1. 30-day?	T	T	T	T	T	T	F	F	T	T	F	F
c2. 31-day?	T	T	F	F	T	T	T	T	T	F	T	F
c3. February?	T	F	T	F	T	F	T	F	T	T	T	T
a1												
Impossible?	Y	Y	Y	N	Y	Y	Y	N	Y	Y	Y	N

Table 3.6   Elimination of Impossible Rules

Conditions	Rule 4	Rule 8	Rule 12
c1. 30-day?	T	F!	F!
c2. 31-day?	F!	T	F!
c3. February?	F!	F!	T
a1			

3.2.3 Redundant and Inconsistent Decision Tables

The ability to recognize (and develop) complete decision tables puts us in a powerful position with respect to redundancy and to inconsistency. The decision table in Table 3.7 is redundant—three conditions and nine rules exist. (Rule 9 is identical to rule 4.) How might this happen? One likely scenario is that it is the result of a maintenance action performed by a less-than-capable developer.

Table 3.7   A Redundant Decision Table

Stub	Rules 1–4	Rule 5	Rule 6	Rule 7	Rule 8	Rule 9
c1	T	F	F	F	F	T
c2	—	T	T	F	F	F
c3	—	T	F	T	F	F
a1	X	X	X	—	—	X
a2	—	X	X	X	—	—
a3	X	—	X	X	X	X

Table 3.8 An Inconsistent Decision Table

Stub	Rules 1–4	Rule 5	Rule 6	Rule 7	Rule 8	Rule 9
c1	T	F	F	F	F	T
c2	–	T	T	F	F	F
c3	–	T	F	T	F	F
a1	X	X	X	–	–	–
a2	–	X	X	X	–	X
a3	X	–	X	X	X	–

Notice that the action entries in rule 9 are identical to those in rules 1–4. As long as the actions in a redundant rule are identical to the corresponding part of the decision table, we do not have much of a problem. If the action entries are different, as they are in Table 3.8, we have a bigger problem.

If the decision table in Table 3.8 were to process a transaction in which $c1$ is true and both c2 and c3 are false, both rules 4 and 9 apply. We can make the following two observations:

1. Rules 4 and 9 are inconsistent.
2. The decision table is nondeterministic.

Rules 4 and 9 are inconsistent because the action sets are different. The whole table is nondeterministic because there is no way to decide whether to apply rule 4 or rule 9.

3.2.4 Decision Table Engines

Even though decision tables are deliberately declarative, they have a structure sufficiently rigorous to support a decision table engine. The input to such a decision table execution engine would be all the information needed to complete the condition entries of a rule. The actions in that rule are then the ones that would be executed. One problem: there is no easy way to control the sequence of executed actions. This could be resolved by using integer entries instead of simple Xs to indicate the execution order of actions in a rule. Alternatively, a decision table engine could be interactive, in the sense that the user would provide sets of values for each rule in the decision table.

3.3 Examples

3.3.1 The NextDate Function

Because of the interesting logical dependencies among its input variables, the NextDate function is popular in software testing circles. NextDate is a function of three variables: month, day, and year. On execution, it returns the month, day, and year of the next day. The variables are all bounded, positive integers, and the boundaries on the year variable are arbitrary.

$$1 <= month <= 12$$
$$1 <= day <= 31$$
$$1801 <= year <= 2100$$

To cast NextDate as a decision table, we will use well-chosen equivalence classes as conditions. (For a more complete discussion of this example, see [Jorgensen 2009].)

$M1 = \{month: month \ has \ 30 \ days\}$
$M2 = \{month: month \ has \ 31 \ days \ except \ December\}$
$M3 = \{month: month \ is \ December\}$
$M4 = \{month: month \ is \ February\}$
$D1 = \{day: 1 \le day \le 27\}$
$D2 = \{day: day = 28\}$
$D3 = \{day: day = 29\}$
$D4 = \{day: day = 30\}$
$D5 = \{day: day = 31\}$
$Y1 = \{year: year \ is \ a \ leap \ year\}$
$Y2 = \{year: year \ is \ a \ common \ year\}$

The Cartesian product of these classes contains 40 elements, so as a starting point, we would expect a decision table with 40 rules. Many of these can be reduced, as seen in Tables 3.9 and 3.10.

Table 3.9 Full NextDate Decision Table (Part 1)

Conditions	1	2	3	4	5	6	7	8	9	10
c1. month in?	M1	M1	M1	M1	M1	M2	M2	M2	M2	M2
c2. day is?	D1	D2	D3	D4	D5	D1	D2	D3	D4	D5
c3. leap year?	–	–	–	–	–	–	–	–	–	–
a1. impossible	–	–	X	X	X	–	–	–	X	X
a2. increment day	X	–	–	–	–	X	X	–	–	–
a3. reset day	–	X	–	–	–	–	–	X	–	–
a4. increment month	–	X	–	–	–	–	–	X	–	–
a5. reset month	–	–	–	–	–	–	–	–	–	–
a6. increment year	–	–	–	–	–	–	–	–	–	–

Table 3.9 Full NextDate Decision Table (Part 2)

Conditions	11	12	13	14	15	16	17	18	19	20	21	22
c1. month in?	M3	M3	M3	M3	M3	M4	M4	M4	M4	M4	M4	M4
c2. day is?	D1	D2	D3	D4	D5	D1	D2	D2	D3	D3	D4	D5
c3. leap year?	–	–	–	–	–	–	Y	N	Y	N	–	–
a1. impossible	–	–	–	–	–	–	–	–	–	X	X	X
a2. increment day	X	X	X	X		X	X	–	–	–	–	–
a3. reset day	–	–	–	–	X	–	–	X	X	–	–	–
a4. increment month	–	–	–	–	–	–	–	X	X	–	–	–
a5. reset month	–	–	–	–	X	–	–	–	–	–	–	–
a6. increment year	–	–	–	–	X	–	–	–	–	–	–	–

Table 3.10   Reduced NextDate Decision Table

Conditions	1–3	4	5	6–9	10	11–14	15	16	17	18	19	20	21, 22
c1. month in?	M1	M1	M1	M2	M2	M3	M3	M4	M4	M4	M4	M4	M4
c2. day is?	D1, D2, D3	D4	D5	D1, D2, D3, D4	D5	D1, D2, D3, D4	D5	D1	D2	D2	D3	D3	D4, D5
c3. leap year?	—	—	—	—	—	—	—	—	Y	N	Y	N	—
a1. impossible	—	—	X	—	—	—	—	—	—	—	—	X	X
a2. increment day	X	—	—	X		X		X	X	—	—	—	—
a3. reset day	—	X	—	—	X	—	X	—	—	X	X	—	—
a4. increment month	—	X	—	—	X	—	—	—	—	X	X	—	—
a5. reset month	—	—	—	—	—	—	X	—	—	—	—	—	—
a6. increment year	—	—	—	—	—	—	X	—	—	—	—	—	—

3.3.2 Windshield Wiper Controller

The windshield wiper is controlled by a lever with a dial on its end. The lever has four positions: OFF, INT (for intermittent), LOW, and HIGH, and the dial has three positions, numbered simply 1, 2, and 3. The dial positions indicate three intermittent speeds, and the dial position is relevant only when the lever is at the INT position. The table below shows the windshield wiper speeds (in wipes per minute) for the lever and dial positions.

c1. Lever	OFF	INT	INT	INT	LOW	HIGH
c2. Dial	n/a	1	2	3	n/a	n/a
a1. Wiper	0	6	12	20	30	60

The problem statement is almost an extended entry decision table as it stands—the only difference here is individual actions for the wiper speeds, as in Table 3.11.

Table 3.11 Windshield Wiper Controller Decision Table

c1. lever at	Off	Intermittent			Low	High
c2. dial at	—	1	2	3	—	—
a1. 0 w.p.m	X	—	—	—	—	—
a2. 6 w.p.m	—	X	—	—	—	—
a3. 12 w.p.m	—	—	X	—	—	—
a4. 20 w.p.m	—	—	—	X	—	—
a5. 30 w.p.m	—	—	—	—	X	—
a6. 60 w.p.m	—	—	—	—	—	X

Conditions c1 and c2 refer to states of the dial and lever; there is no indication of the lever and dial move events. Thus the context-sensitive input events cannot be observed in this decision table. In a sense, the rules in which the lever is in the INT position serve to enable the dial positions—at least, they give meaning to the dial positions. We will have a more explicit model of this in Chapter 10.

3.3.3 Railroad Crossing Gate Controller

In a town in northern Illinois, there is a railroad crossing where Poplar Avenue crosses the Chicago North Western railroad tracks. There are three separate tracks at the intersection, and each track has sensors to detect when a train approaches the intersection and when it leaves. When no trains are present (or approaching), the crossing gate is up. When the first train approaches, the crossing gate is lowered, and when the last train leaves, the gate is raised. If a second or third train arrives while a train is in the intersection, there is no action on the gate (because it is already down).

Port input events	Port output events	Data
p1. train arrival	p3. lower crossing gate	d1. The number of trains in the crossing
p2. train departure	p4. raise crossing gate	

Table 3.12 is an excellent example of a Mixed Entry Decision Table (MEDT). Condition c1 acts as memory, and is updated by actions a4 and a5. The "impossible

Table 3.12   Mixed Entry Decision Table for the Railroad Crossing Gate Controller

Rules	1	2	3	4	5	6	7	8	9	10	11	12	13	14	15	16
c1: Train count is...	0				1				2				3			
c2: Train arrives	T	T	F	F	T	T	F	F	T	T	F	F	T	T	F	F
c3: Train departs	T	F	T	F	T	F	T	F	T	F	T	F	T	F	T	F
a1: Lower gate		x														
a2: Raise gate						x										
a3: Do nothing			x	x			x	x				x				x
a4: Increment train count		x			x				x							
a5: Decrement train count							x					x				x
a6: Impossible	x		x										x	x		

rules" are indicated by entries in action a6. Rules 1 and 3 are impossible because, if there are no trains in the intersection, how could one depart? Rules 13 and 14 are impossible because the three tracks are already occupied. The "Do nothing" actions when c2 and c3 are both true (rules 5 and 9) could be replaced by increment and decrement actions, but the Do nothing action shows how these would cancel each other out. (Also, decision tables do not represent time, so that is another reason for the Do nothing entries.)

3.4 Deriving Test Cases from Decision Tables

Deriving test cases from a decision table is possible, but the process can be awkward. The biggest problem is that decision tables are, or at least should be, declarative; thus order cannot be relied on. This is exacerbated if a decision table is optimized, because the optimization process may permute all or any of the conditions, actions and rules. The sequential nature of a test case is therefore easily lost. This is often compounded by the fact that simplification may obscure interesting test cases.

On the plus side, we can have a deserved sense of completeness if a decision table is developed carefully. Also, as we can recognize redundancy and nondeterminism, the corresponding test cases can avoid these problems. Completeness is limited, however—we can never automatically detect a missing condition, or a missing action. This is where domain experience comes in, and why deriving test cases from a decision table is best done manually. Decision tables can be appropriate for computational applications; less so for event-driven applications. Further, if events and data contexts are carefully separated in the condition stub, it will be easier to identify context-sensitive input events. Rules are good sources for test cases in computational, decision-intensive applications. In event-driven applications, a test case will probably correspond to a sequence of rules.

3.4.1 The Insurance Premium Problem Decision Table

The definition of the Insurance Premium Problem (Chapter 1) supports nearly direct development of a Mixed Entry Decision Table. The age and "at fault claims" variables have defined ranges that lead directly to the extended entry conditions $c1$ and $c2$ in Table 3.13. The premium reductions are both boolean, so conditions $c3$ and $c4$ are limited entry conditions. For space reasons, Table 3.13 is split into four parts. There are no impossible rules, nor are there any Do Nothing actions. The conditions where insurance is denied are in Table 3.13, Part 4.

As would be true for any application, the possibility of developing a complete decision table is a major benefit—we know that nothing is missing. The declarative property of decision tables is potentially problematic here. The age-based multiplier should be applied before adding the penalties due to "at fault" claims; otherwise the claims penalty could be unduly increased. There is nothing to prevent this if the absolute declarative nature of decision tables is used to generate test cases—this is where domain experience is needed.

There are 39 distinct rules in the MEDT for the Insurance Premium Problem. Each rule defines an abstract test case, and they correspond to the paths in the Insurance Premium Flowchart in Chapter 2. Reducing these to concrete test cases cannot be derived from the decision table. These 39 test cases align closely with those derived from the flowchart in Chapter 2, except for the compound condition age <16 or age >90. Table 3.14 can be derived from Table 3.13 (Part 1).

Table 3.15 cannot be derived from Table 3.13, but it is easily developed (manually) from the rules in Table 3.13.

Table 3.13   Insurance Premium Problem Decision Table (Part 1)

c1. Age	16 <= age < 25											
c2. At fault claims	0				1 to 3				4 to 10			
c3. goodStudent?	T	T	F	F	T	T	F	F	T	T	F	F
c4. nonDrinker?	T	F	T	F	T	F	T	F	T	F	T	F
a1. Base times 1.5	x	x	x	x	x	x	x	x	x	x	x	x
a2. Base times 1												
a3. Base times 1.2												
a4. add $0 to Base	x	x	x	x								
a5. add $100 to Base					x	x	x	x				
a6. add $300 to Base									x	x	x	x
a7. subtract $50 from Base	x	x			x	x			x	x		
a8. subtract $75 from Base	x		x		x		x		x		x	
Rule	1	2	3	4	5	6	7	8	9	10	11	12

Table 3.13   Insurance Premium Problem Decision Table (Part 2)

c1. Age	25 <= age < 65											
c2. At fault claims	0				1 to 3				4 to 10			
c3. goodStudent?	T	T	F	F	T	T	F	F	T	T	F	F
c4. nonDrinker?	T	F	T	F	T	F	T	F	T	F	T	F
a1. Base times 1.5												
a2. Base times 1	x	x	x	x	x	x	x	x	x	x	x	x
a3. Base times 1.2												
a4. add $0 to Base	x	x	x	x								
a5. add $100 to Base					x	x	x	x				
a6. add $300 to Base									x	x	x	x
a7. subtract $50 from Base	x	x			x	x			x	x		
a8. subtract $75 from Base	x		x		x		x		x		x	
Rule	13	14	15	16	17	18	19	20	21	22	23	24

Table 3.13 Insurance Premium Problem Decision Table (Part 3)

c1. Age	65 <= age <= 90											
c2. At fault claims	0				1 to 3				4 to 10			
c3. goodStudent?	T	T	F	F	T	T	F	F	T	T	F	F
c4. nonDrinker?	T	F	T	F	T	F	T	F	T	F	T	F
a1. Base times 1.5												
a2. Base times 1												
a3. Base times 1.2	x	x	x	x	x	x	x	x	x	x	x	x
a4. add $0 to Base	x	x	x	x								
a5. add $100 to Base					x	x	x	x				
a6. add $300 to Base									x	x	x	x
a7. subtract $50 from Base	x	x			x	x			x	x		
a8. subtract $75 from Base	x		x		x		x		x		x	
Rule	25	26	27	28	29	30	31	32	33	34	35	36

Table 3.13 Insurance Premium Problem Decision Table (Part 4)

c1. Age	<16	>90	—
c2. At fault claims	—	—	>10
c3. goodStudent?	—	—	—
c4. nonDrinker?	—	—	—
a9. cannot be insured	x	x	x
Rule	37	38	39

3.4.2 The Garage Door Controller Decision Table

Table 3.16 is the decision table model for the Garage Door Controller. It shows exactly which events occur in the possible given contexts of the garage door (entries to condition c1). It is split into two parts for space reasons: Part 1 deals with closing

Table 3.14   Insurance Premium Abstract Test Case

Insurance Premium Abstract Test Case (from Rule 1)	
Preconditions	Base, Age, Claims, goodStudent, and nonDrinker are known
Inputs	Actions
1. Age in 16 <= age < 25	2. Multiply Base by 1.5
3. Claims = 0	4. Multiply Base by 1.0
5. goodStudent is True	6. Subtract $50 from Base
7. nonDrinker is True	8. Subtract $75 from Base
Postconditions	Base contains the premium value

Table 3.15   Insurance Premium Concrete (Real) Test Case

Insurance Premium Real Test Case (from Rule 1)	
Preconditions	Base = $600
Inputs	Actions
1. age = 20	2. $900 = $600 * 1.5
3. claims = 0	4. $900 = $900 − $0
5. goodStudent is True	6. $850 = $900 − $50
7. nonDrinker is True	8. $775 = $850 − $75
Postconditions	Premium = $775

the garage door; part 2 deals with opening the door. We can also see the presence of context-dependent input events—for example, in rule 1, the response to a control signal is to start the drive motor in the down direction, and in rule 5, the motor is stopped for the same input event. Table 3.16 uses the "F!" (must be false) notation to show the proscribed events in each context (some of which are indicated by the impossible rules). This also follows a common modeling convention that prohibits simultaneous events. As this table always allows every event in each context, we have several impossible rules.

The easy answer for deriving test cases is that a rule is/should be a test case. The problem is, the rules are very short, only involving one input event. As such, they are better understood as stimulus/response pairs. In Chapter 4, we shall see that these

Table 3.16 Garage Door Controller Decision Table (Part 1)

Rule	1	2	3	4	5	6	7	8	9	10	11	12
c1. Door is	Up				Closing				Stopped Going Down			
c2. control device signal	T	F!	F!	F!	T	F!	F!	F!	T	F!	F!	F!
c3. end of down track reached	F!	T	F!	F!	F!	T	F!	F!	F!	T	F!	F!
c4: end of up track reached	F!	F!	T	F!	F!	F!	T	F!	F!	F!	T	F!
c5. light beam interruption	F!	F!	F!	T	F!	F!	F!	T	F!	F!	F!	T
a1. start drive motor down	x								x			
a2. start drive motor up												
a3. stop drive motor					x	x						
a4. reverse motor down to up								x				
a5. do nothing				x								x
a6. impossible		x	x				x			x	x	
a7. repeat table	x			x	x	x		x	x			x

Table 3.16 Garage Door Controller Decision Table (Part 2)

Rule	13	14	15	16	17	18	19	20	21	22	23	24
c1. Door is	Down				Opening				Stopped Going Up			
c2. control device signal	T	F!	F!	F!	T	F!	F!	F!	T	F!	F!	F!
c3. end of down track reached	F!	T	F!	F!	F!	T	F!	F!	F!	T	F!	F!
c4. end of up track reached	F!	F!	T	F!	F!	F!	T	F!	F!	F!	T	F!
c5. light beam interruption	F!	F!	F!	T	F!	F!	F!	T	F!	F!	F!	T
a1. start drive motor down												
a2. start drive motor up	x								x			
a3. stop drive motor					x		x					
a4. reverse motor down to up												
a5. do nothing				x				x				x
a6. impossible		x	x			x				x	x	
a7. repeat table	x			x	x		x	x	x			x

rule-based stimulus/response pairs align nicely with transitions in the finite state machine (FSM) description of the Garage Door Controller. Deriving "full" test cases from Table 3.16 is possible, but only manually. Domain experience is required to deduce sensible sequences of rules that are only hinted at by action a7, the Repeat Table action. The declarative property is another complicating factor—there are some "natural" orders of rule selection and execution, but these are not expressed. Another problem is that the decision table is a "one time" description. There is no way to show the possible cycles of stopping and restarting a door in motion unless the repeat table action is selected.

3.4.3 Test Cases for the Garage Door Controller

On the surface, the 24 rules in the Garage Door Controller, MEDT would suggest 24 test cases. The five "do nothing" actions (Rules 4, 12, 16, 20, and 24) are curious—how do you test for nothing happening. Each of these rules involves the light beam interruption event, but that event is only active when the door is closing (rule 8). These five "do nothing" rules are examples of a physically possible event that is ignored. (Later, in Chapter 8, we will be able to take a much closer look at how the light beam sensor is enabled and disabled.) The ten impossible cases (Rules 2, 3, 7, 10, 11, 14, 15, 18, 22, and 23) are physically impossible; they refer to the end-of-track events when they cannot possibly occur. We are left with nine "actual" test cases (Rules 1, 5, 6, 8, 9, 13, 17, 19, and 21). As the actions are all legitimate port output events, these nine are concrete test cases. As previously noted, they are short stimulus/response pairs. We can connect these rule-based, short test cases into rule sequences that are more similar to end-to-end test cases. The five test cases from Chapter 2 (Table 2.3) are shown in Table 3.17 as sequences of rules.

Table 3.17 Rule Sequences for Garage Door Test Cases

Sample Garage Door Decision Table Rule Sequences		
Path	Description	Rule Sequence
1	Normal door close	1, 6
2	Normal door close with one intermediate stop	1, 5, 9, 6
3	Normal door open	13, 17
4	Normal door open with one intermediate stop	13, 17, 21, 19
5	Closing door with light beam interruption	1, 8, 19

3.5 Advantages and Limitations

Decision tables are clearly the model of choice for logic-intensive applications. This is particularly true when conditions refer to equivalence classes that have dependencies, as in NextDate. Also, the ability to algebraically simplify decision tables results in an elegant approach to minimization. Calculations can only be expressed as actions, but the absence of order is problematic. Finally, input events must be represented as conditions, and output events as actions, but again, the lack of order makes this approach inadequate for event-driven applications. Table 3.18 summarizes the behavioral issue representation of decision tables.

3.6 Lessons Learned from Experience

I was a member of the CODASYL Decision Table Task Group during the 1970s. We were tasked with creating a "definitive" description of decision tables, including recommended best practices. The final product was published by the Association for Computing Machinery (ACM) in an enlarged, paperback edition. Along the way, we had several interesting episodes. One was when the task group member from Belgium, Maurice Verhelst, proposed that we make a decision table for recently enacted rent control laws that were very confusing. At the following quarterly meeting, we merged individual submissions into a solid decision table that clarified many of the issues, including a couple of inconsistencies.

Another task group member, Lewis Reinwald, had a decision table-based program to clean up FORTRAN programs that had been heavily modified, usually by a variety of developers who either did not understand the original program or the many previous changes to it. In a test case, Mr. Reinwald took a large, heavily maintained, (nightmare) FORTRAN program of more than 2000 source statements, simplified the decision table derived from the source code, "cleaned up" the decision table manually, and then generated an improved version of the original program. The improved version was on the order of 800 statements.

From 1978 to 1981, I worked for the Italian subsidiary of my employer. As part of the requirement specification process, we used a combination of finite state machines that were further detailed by decision tables. We repeatedly found that the structure of the decision tables forced consideration of situations that otherwise would not have been recognized until very late in the development.

In my university classes, I used an example of teacher retirement incentives that was passed by the state legislature. It contains a pair of dependent conditions and results in an ambiguous definition of retirement benefit computation. The common thread among these three vignettes is that decision tables can be very effective at clarifying confusing business rules.

Table 3.18   Representation of Behavioral Issues with Decision Tables

Issue	Represented?	Example
Sequence	No	Decision tables are/should be declarative
Selection	Yes	The whole point of decision tables!
Repetition	Yes	Must use a repeat table action
Enable	Indirectly	An enabling condition in conjunction with remaining enabled conditions
Disable	Indirectly	A disabling condition in conjunction with remaining enabled conditions
Trigger	No	
Activate	Indirectly	Only in the sense of successive enabling and disabling rules, BUT there is no sense of order, so this is VERY tenuous
Suspend	Indirectly	A true suspending condition in combination with the remaining condition entries either don't care or F!
Resume	No	
Pause	No	
Conflict	No	
Priority	No	
Mutual exclusion	Yes	Rules are mutually exclusive
Concurrent execution	No	
Deadlock	No	
Context-sensitive input events	Yes	Event is one condition, contexts are separate conditions
Multiple-context output events	Yes	Same action(s) in different rules
Asynchronous events	No	
Event quiescence	No	

References

[CODASYL 1978]
CODASYL Systems Group, *DETAB-X, Preliminary Specification for a Decision Table Structured Language*, Association for Computing Machinery, New York, 1978.
[Jorgensen 2009]
Jorgensen, Paul C., *Modeling Software Behavior—A Craftsman's Approach*. CRC Press, Boca Raton, FL, 2009.

Chapter 4

Finite State Machines

Finite state machines (FSMs) have become a fairly standard notation for requirements specification. All the real-time extensions of Structured Analysis use some form of FSM, and nearly all forms of object-oriented analysis require either FSMs, or their extension, Statecharts, the behavioral model of choice for UML. This chapter is particularly important for model-based testing (MBT) because most of the commercial and open source MBT tools are designed for FSMs. (Much of Section 4.1 is adapted from *Modeling Software Behavior—A Craftsman's Approach* [Jorgensen 2009].)

4.1 Definition and Notation

A finite state machine (FSM) is a directed graph in which nodes are states and edges are transitions. Source and sink nodes become initial and terminal states, paths in the finite state machine are modeled as directed graph paths, and so on. Most finite state machine notations add information to the edges (transitions) to indicate the cause of the transition and actions that occur as a result of the transition. In some circles, this is referred to as an Extended Finite State Machine (EFSM). Here is a formal definition.

Definition: A *Finite State Machine* is a four-tuple (S, T, In, Out)

where:
 S is a set of states
 T is a set of transitions
 In is a set of inputs that cause transitions
 Out is a set of outputs that can occur on transitions

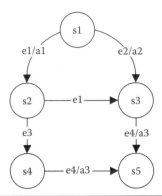

Figure 4.1   Finite state machine for discussion purposes.

In the finite state machine in Figure 4.1, we have the following values for the four-tuple:

S = {s1, s2, s3, s4, s5}
T = {<s1, s2>, {<s1, s3>, {<s2, s3>, <s2, s4>, <s3, s5>, <s4, s5>}
In = {e1, e2, e3, e4}
Out = {a1, a2, a3}

This machine contains five states and six transitions, which are shown, respectively, as circles and edges. The labels on the transitions follow a convention that the "numerator" is the event that causes the transition, and the "denominator" is the action that is associated with the transition. Inputs can be either events, data conditions, or a logical combination of events and data conditions. The transition from s1 to s2, for example, is caused by the input event e1 and generates the output event a1. Event e1 is a context-sensitive input event—the response depends on the context, expressed in this model as a state in which the event occurs. The inputs are mandatory—transitions do not just happen, they are caused by inputs; but the output actions are optional.

Paths through a finite state machine can be identified in three ways: as a sequence of states, as a sequence of state transitions, or as a sequence of inputs that cause transitions. A given finite state machine may represent a large number of distinct paths, much like the intension of a database represents many possible extensions (distinct populations). In some situations, it is necessary to identify an initial state; this may be obvious if there are no transitions into a state. Symmetrically, there may be at least one, but possibly a plurality of terminal states. Again, these are easily identified as states with no exiting transitions. If a finite state machine is strongly connected (as a directed graph), then it is necessary to add a "transition from nowhere" to show an intended source state as we shall see with the Crossing Gate Controller (Figure 4.7) and the Garage Door Controller finite state machine (Figure 4.13).

4.1.1 *Matrix Representations of Finite State Machines*

There are two common textual representations of a finite state machine—a Transition Table and an Event Table; these are shown, respectively, in Tables 4.1 and 4.2 for the example finite state machine in Figure 4.1.

If a row contains no entries, the corresponding state is a terminal state. Symmetrically, if a state never appears in a column, for example state s1 in Table 4.1, it is an initial state.

The information in Tables 4.1 and 4.2 is both necessary and sufficient to recreate a finite state machine drawing such as the one in Figure 4.1. (Obviously, only a topologically equivalent drawing can be generated. There is no "placement" information in either table. This has implications for MBT tools.)

Since finite state machines are directed graphs, two matrices from directed graph theory also apply—the Adjacency Matrix (in Table 4.3) and the Reachability Matrix (Table 4.4).

Definition: For a directed graph G consisting of *n* nodes, its *adjacency matrix* is the matrix $A = (a_{i,j})$, where the element in row *i*, column *j*, is 1 if and only if there is an edge from node *i* to node *j*.

Table 4.1   Transition Table for Figure 4.1

States\Events	e1	e2	e3	e4
s1	s2	s3	—	—
s2	s3	—	s4	—
s3	—	—	—	s5
s4	—	—	—	s5
s5	—	—	—	—

Table 4.2   Event Table for Figure 4.1

States\Events	e1	e2	e3	e4
s1	a1	a2	—	—
s2	—	—	—	—
s3	—	—	—	a3
s4	—	—	—	a3
s5	—	—	—	—

Table 4.3   Adjacency Matrix for Figure 4.1

	s1	s2	s3	s4	s5
s1	0	1	1	0	0
s2	0	0	1	1	0
s3	0	0	0	0	1
s4	0	0	0	0	1
s5	0	0	0	0	0

Table 4.4   Reachability Matrix for Figure 4.1

	s1	s2	s3	s4	s5
s1	1	1	1	1	1
s2	0	1	1	1	1
s3	0	0	1	0	1
s4	0	0	0	1	1
s5	0	0	0	0	1

Definition: For a directed graph G consisting of *n* nodes, its *reachability matrix* is the matrix $R = (r_{i,j})$, where the element in row *i*, column *j*, is 1, if and only if there is a path from node *i* to node *j*.

The reachability matrix R of directed graph G can be computed from the adjacency matrix as follows:

$$R = I + A + A^2 + \ldots + A^k$$

where k is the longest path in G, I is the identity matrix, and in the matrix multiplication, $1 + 1 = 1$. Each power of the adjacency matrix describes the paths of the length of the exponent. The identity matrix, I, shows the paths consisting of 0 edges, A shows the paths of length 1, and so on.

4.1.2 Textual Representations of Finite State Machines

Although the matrix representations are well known, they are not as helpful as more textual forms. Table 4.5 compresses the Event and Transition tables into what is perhaps a more useful form.

Table 4.5   Alternative Representation for Figure 4.1

Transition	From State	To State	Cause	Output
1	s1	s2	e1	a1
2	s1	s3	e2	a2
3	s2	s3	e1	—
4	s2	s4	e3	—
5	s3	s5	e4	a3
6	s4	s5	e4	a3

Figure 4.1 can also be represented in a template, as shown below.

State <state Id> <state name> has transitions
  <transition 1> To <to state> caused by:
    <input(s)> and generates:
    <output(s)>

Here is the template completed for state s1 in Figure 4.1:

State s1 < > has transitions
  transition 1 To s2 caused by:
    e1 and generates a1
  transition 2 To s3 caused by:
    e2 and generates a2

4.1.3 Conventions and Restrictions of Finite State Machines

There are two major restrictions on finite state machines—they have no memory, and the states must be independent. In addition, there are three other conventions, explained in this subsection.

Restriction 1: There is no memory in a finite state machine.

As software developers, we tend to think in terms of memory as part of some algorithm we are implementing. It is a fundamental assumption in flowcharts, as we have seen in Chapter 2. Figure 4.2 illustrates a violation of the no memory restriction. The "history" of what states a path that has traversed cannot affect transitions from a "later" state.

Restriction 2: States in a finite state machine are independent.

Memory and independence can be confused. States s5 and s6 in Figure 4.2 are clearly dependent on a history that traverses one (or both) of states s2 and s3. When

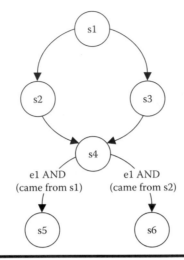

Figure 4.2   Transitions that use memory.

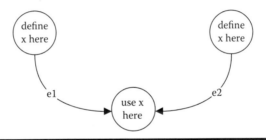

Figure 4.3   Dependencies in a finite state machine.

there are internal actions associated (but not visible in) with states, these actions must not affect "later" states. The most common dependence is via global variables, or possibly a state that establishes a configuration that cannot be handled by a later state. Another possibility is that an output action on a transition from one state may affect a subsequent state. As a generalization, when there are dependencies among states, paths that traverse dependent states are frequently impossible.

Convention 1: There is only one "active" (some authors prefer "current") state at any point (in time).

Conversationally, we speak of "being in a certain state." To clarify this, we can think of a state as a proposition that is True during some interval of time. A state transition then is seen as something that makes one state proposition False, and a subsequent state proposition True. With this view, this convention means that the state propositions are mutually exclusive. If a given finite state machine has an initial state, that is the default current state.

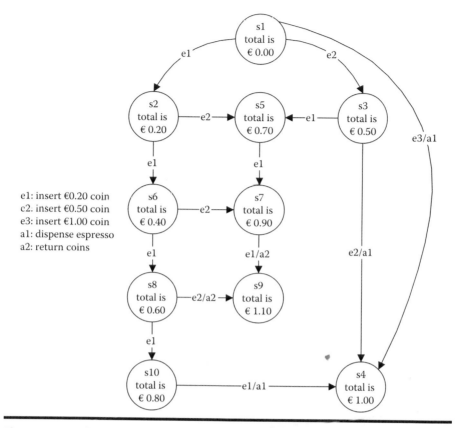

Figure 4.4 **A finite state machine for an espresso vending machine.**

Definition: A *state* in a finite state machine is an interval of (execution) time during which a certain proposition is true.

Figure 4.4 is a finite state machine for a vending machine that dispenses espresso coffee. A single espresso costs 1 Euro (€1.00), and the vending machine can only accept €0.20, €0.50, and €1.00 coins. The state propositions are shown in the figure. The initial state proposition, total is €0.00, is changed when one of the coin insertion events occur. If a €0.20 coin is inserted, the new state proposition total is €0.20. Figure 4.4 also shows how state names (and the corresponding state propositions) simulate memory. Also notice that s1 is the initial state, and s4 and s9 are terminal states.

Convention 2: There are no simultaneous events in a finite state machine.

This convention preserves the single active (current) state property. It also preserves sanity—what would happen in the espresso vending finite state machine if, when s1 is the current state, events e1 and e2 were truly simultaneous. What would be the next state? There is one last restriction.

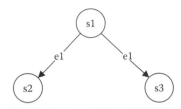

Figure 4.5 A nondeterministic finite state machine.

Convention 3: A nondeterministic finite state machine occurs when a state has transitions to two (or more) subsequent states caused by the same event (see Figure 4.5).

4.2 Technique

In his famous paper, "No Silver Bullets" [Brooks 1986], Fred Brooks separates the difficulties of software development into two groups, which he named *essence* and *accident*. The essential part of software development is the inherent complexity and other constraints that cannot be avoided. The accidental part has more to do with choices developers can/do make. As with the other models, the definitions and notation are the accidental part; using a model well is the essence.

4.2.1 Interpretations of Finite State Machine Elements

Knowing how to use finite state machines well is what makes a modeler a craftsperson. The notational tools available to one who models an application with a finite state machine are the states themselves, the inputs that cause state transitions, and actions that may occur or with a transition.

States are commonly used to represent any of the following:

- Stages of processing
- Data conditions
- Consequences of input events that have occurred
- Hardware configurations
- Device status

Recall that we can usefully think of a state as a proposition that is either True or False. Each of the listed interpretations of a state makes sense as a proposition that is true. The states in Figure 4.12 (the Insurance Premium Problem) are good examples of states as stages of processing. Another common example is the use of states to represent stages of processing, such as a response to a field trouble report: Received, Analyzed, Repaired, Tested, Released. The states in Figure 4.4 (the espresso vending

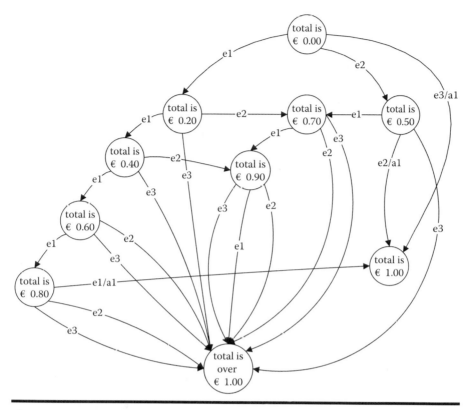

Figure 4.6 Vending Machine finite state machine with pre and proscribed behaviors.

machine) are all examples of using a state to represent a data condition. The states in Figure 4.6 show the consequences of input events that occur.

Inputs that cause state transitions can be of any of these: port events, data conditions, passage of time, or a logical combination. Actions that are associated with transitions can be of any of these: port output events, changes in data values, internal system actions/functions, or a logical combination. If the inputs and actions in a finite state machine are all port input and port output events, as they are in Figures 4.6 and 4.7, information on a path translates almost directly to a system level test case (Table 4.6).

4.2.2 Selected Finite State Machine Practices

4.2.2.1 The State Explosion

One consequence of Restriction 2 (states must be independent) is that, when independent states are merged into a finite state machine, the famous "state explosion" occurs. In the Windshield Wiper Controller discussion (Section 4.3.2), the Lever finite state

Table 4.6   A Use Case Derived from the Windshield Wiper Controller Finite State Machine (Figure 4.9)

Use Case Name:	Windshield Wiper Controller Execution Table Example	
Use Case ID:	WWUC-1	
Description:	Operator moves through a typical sequence of lever and dial events.	
Preconditions:	1. Lever position is OFF, Dial position is 1.	
Event Sequence:	**Input Event**	**System Response**
	1. e1: move lever up one position	2. a2: deliver 6 wipes per minute
	3. e3: move dial up one position	4. a3: deliver 12 wipes per minute
	5. e1: move lever up one position	6. a5: deliver 30 wipes per minute
	7. e1: move lever up one position	8. a6: deliver 60 wipes per minute
	9. e2: move lever down one position	10. a5: deliver 60 wipes per minute
	11. e3: move dial up one position	(no effect on wiper speed)
	12. e2: move lever down one position	13. a4: deliver 20 wipes per minute
	14. e2: move lever down one position	15. a1: deliver 0 wipes per minute
Postconditions:	1. Lever position is OFF, Dial position is 3	

machine and the Dial finite state machine are separate machines because they are physically independent devices (with intended logical dependencies). Figure 4.9 is the result of making the cross product of the two machines. Making a cross product is always assumed to operate on independent items. It is a multiplicative process. A former graduate student of mine at Arizona State University, Tempe, Arizona once developed a generalized formula for how many states in a finite state machine of an elevator system with n floors and m elevator cars. For a building with 9 floors and 3 elevators (the one I teach in), the full cross product has 110,596 states. The best answer for the state explosion is to move to Statecharts. (David Harel made a Statechart model of the elevator system that fits on only three pages.)

4.2.2.2 Prescribed and Proscribed Behaviors

When developing a finite state machine, we tend to think only of the prescribed (intended) behaviors. It is usually more difficult to come up with the proscribed (prohibited) behaviors. This can be important when the inputs are physical events over which the system being described has no control. Figure 4.6 shows all possible input events that can physically occur in Figure 4.4. We will revisit this in Chapter 7.

4.2.3 Finite State Machine Engines

Finite state machines can be executed, but the conventions described in Section 4.1 must apply. If a finite state machine contains loops as in the Railroad Crossing Gate Controller example (Figure 4.11), there is a countably infinite number of paths. We sometimes speak of a path in a FSM as a scenario; alternatively, paths can correspond to use cases, or even the user stories of agile development. In the Railroad Crossing Gate Controller, the scenario "A train arrives, followed by a second train. A train departs, then the other train departs" has the state sequence <s1, s2, s3, s2, s1> and the event sequence <p1, p1, p2, p2>.

Recall that the basic structure of a finite state machine consists of (S, T, In, Out) where:
S is a set of states,
T is a set of transitions,
In is a set of inputs that cause transitions,
Out is a set of outputs that can occur on transitions, and optionally,
One state is designated as the initial (or source) state.

Each of these elements is needed in a finite state machine execution engine. A likely implementation consists of a graphical user interface in which the set of states is given, and the set of events (and other conditions, if needed) is available, probably via a drop-down menu. Once an initial state is identified (interactively, by the customer/user), the user can "cause" an event to occur, thereby generating a transition to the next state. If actions occur on the transition, they are noted. The customer/user can move from state to state by selecting events. One interesting possibility is that the user is free to select events that correspond to proscribed behavior, such as paying too much for an espresso. A Finite State Machine engine would produce an execution table similar to the one in Table 4.7 for the typical use case after (Table 4.6).

Table 4.7 Execution Table for the Windshield Wiper Controller

Step	In States	Input Event	Next States	Output Action (wpm = wipes per minute)
1	Off, 1	e1: move lever up one position	Int, 1	a2: deliver 6 wpm
2	Int, 1	e3: move dial up one position	Int, 2	a3: deliver 10 wpm
3	Int, 2	e1: move lever up one position	Low, 2	a5: deliver 30 wpm
4	Low, 2	e1: move lever up one position	High, 2	a6: deliver 60 wpm
5	High, 2	e2: move lever down one position	Low, 2	a5: deliver 30 wpm
6	Low, 2	e3: move dial up one position	Low, 3	a5: deliver 30 wpm
7	Low, 3	e2: move lever down one position	Int, 3	a4: deliver 20 wpm
8	Int, 3	e2: move lever down one position	Off, 3	a1: deliver 0 wpm

4.3 Examples

4.3.1 Windshield Wiper Controller

There are two independent devices in the Windshield Wiper Controller—the lever and the dial. The input and output events are as follows:

Input Events	Output Events	
e1: move lever up one position	a1: deliver 0 wipes per minute	a5: deliver 30 wipes per minute
e2: move lever down one position	a2: deliver 6 wipes per minute	a6: deliver 60 wipes per minute
e3: move dial up one position	a3: deliver 10 wipes per minute	
e4: move dial down one position	a4: deliver 20 wipes per minute	

The two device-oriented finite state machines are shown in Figure 4.7. Notice the question mark actions on every transition in the Dial finite state machine and on the transitions to state s2 in the Lever finite state machine. These actions are all undetermined because neither of the finite state machine knows the active state of the other. Second, event e1 is a context-sensitive input event—the response is different depending on the context in which event e1 occurs.

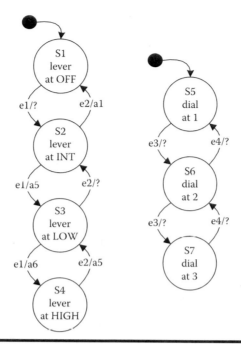

Figure 4.7 **Lever and Dial finite state machines.**

The lever events are a good example of a subtlety about event choices. We could replace event e1: move lever up one position with three more specific events (and similar extensions for the other events):

e.1.1: move lever Off to Int
e.1.2: move lever Int to Low
e.1.3: move lever Low to High

This choice is tantamount to moving the context of a context-sensitive input event into a set of more specific input events. In a similar way, multiple context output events can be made more specific. The choice between context sensitivity and more specific events is really one of the styles. If the state machines are used to automatically generate test cases, the more specific version of events is preferred.

Our next step is to remove the ambiguity of transition actions. Figure 4.8 shows a brute force way—simply take the Cartesian product of the two finite state machines. This process leads to the famous "finite state machine explosion." There is much redundancy in Figure 4.8: the triples of horizontal states (in which the lever position is constant) are only interesting when the lever is in the INT (intermittent) state. This is not nearly as elegant as the extended entry decision table (EEDT) with the Don't Care entries in Table 3.10 (Chapter 3), and it certainly will not scale up well for devices with more states.

Figure 4.9 almost resolves the ambiguity we had in Figure 4.8, but there is still no information about which transition occurs "in" the Lever at INT state. We could use the

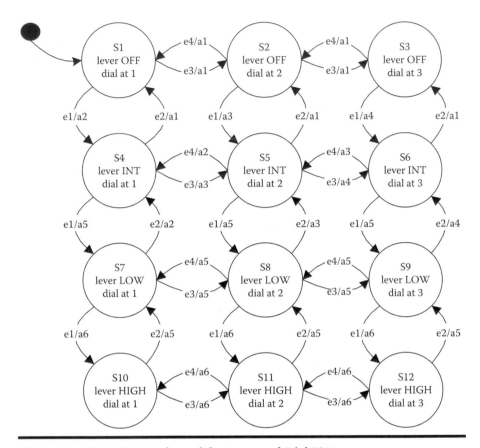

Figure 4.8   Cartesian product of the Lever and Dial FSMs.

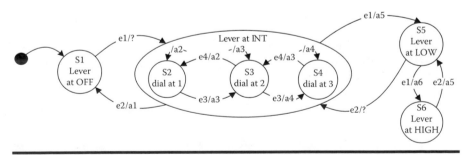

Figure 4.9   Hierarchical FSM for the Windshield Wiper.

mechanism of communicating finite state machines, sometimes abbreviated as CFMs. Full treatment of that involves more formalism than most of us need, so for now, we just use the notion of a message. (Use of messages in CFMs probably predates the use of messages in object-oriented programming.) Rather than a true output action on a transition, a CFM can just send a message to another finite state machine as in Figure 4.10.

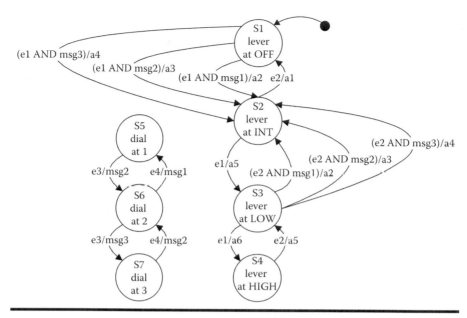

Figure 4.10 **Lever and Dial finite state machines as communicating finite state machines.**

(We shall see exactly this mechanism in Statecharts.) The messages from the Dial finite state machine to the Lever finite state machine just provide the position of the Dial (1, 2, or 3). In Figure 4.10, the messages are: msg1: dial at 1, msg2: dial at 2, msg3: dial at 3.

The only remaining question has to do with event quiescence: What if no event occurs in the Dial finite state machine? If no dial event occurs, no message will be sent to the Lever finite state machine. Thus once the Lever finite state machine is in either state s1 or s3, it is stuck there until a dial event occurs. The actual answer in "pure communicating finite state machines" is that, because both finite state machines are communication, each always knows the active state of the other. We will see an even clearer expression of this in Chapter 7 on Statecharts. In fact, the transitions to states s2, s3, and s4 in Figure 4.9 borrow some transition notation from Statecharts.

4.3.2 Railroad Crossing Gate Controller

In a town in northern Illinois, there is a railroad crossing where Poplar Avenue crosses the Chicago North Western railroad tracks. There are three separate tracks at the intersection, and each track has sensors to detect when a train approaches the intersection and when one leaves. When no trains are present (or approaching) the crossing gate is up. When the first train approaches, the crossing gate is lowered, and when the last train leaves, the gate is raised. If a second or third train arrives while a train is in the intersection, there is no action on the gate (because it is already down). The Railroad Crossing Gate Controller has two input events and two output events. We can assume that train arrivals and departures are sensed in a timely way so no safety issues need to be considered. Here we take a closer look at the Railroad Crossing Gate Controller. The port input and port output events from the decision table formulation in Chapter 3 are

Port Input Events	Port Output Events	Data
p1. train arrival	p3. lower crossing gate	d1. the number of trains in the crossing
p2. train departure	p4. raise crossing gate	

As a finite state machine, we use states to act as the memory of how many trains are in the crossing. The four states replace the data condition of the decision table formulation (see Figure 4.11).

Refer to Figure 4.7 for the finite state machine of the Railroad Crossing Gate Controller. The states act as memory, and they align exactly with the MEDT (Mixed Entry Decision Table) formulation (Table 3.10) in Chapter 3. As a decision table, we used impossible rules to refer to proscribed possibilities, for example, how could a train depart when there are no trains in the intersection? In Chapter 3, there are 16 rules in the Mixed Entry Decision Table for the Railroad Crossing Gate Controller. Four of these are noted as impossible: Rules 1 and 3 refer to a train departure from an empty intersection. Rules 13 and 14 refer to train arrivals at an already full intersection. (We would hope that this could not happen...) One could argue that Rule 13 should be a "do nothing" rule, not an impossible one, with the understanding that as one train departed, the arriving one took its place in the intersection. BUT this presumes an order on the conditions, which contradicts the declarative nature of decision tables. Rules 5 and 9 also refer to "simultaneous" train arrivals and departures, but they occur when there is at least one unoccupied track. In a sense, they "cancel each other out" because the net effect is that the train count (State) is unchanged.

The finite state machine in Figure 4.11 removes these awkward possibilities by only showing prescribed behaviors. Table 4.8 maps the rules from the Mixed Entry Decision Table for the Railroad Crossing Gate Controller to both state sequences and event sequences in the Railroad Crossing Gate Controller finite state machine.

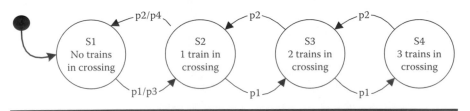

Figure 4.11   Finite State Machine for the Railroad Crossing Gate Controller.

Table 4.8 Mapping the Railroad Crossing Controller Decision Table with Its Finite State Machine Model

MEDT Rule	FSM State Sequence	FSM Event Sequence
1	impossible	
2	s1, s2	p1, p3
3	impossible	
4	do nothing	null, null
5	Cannot show. Depends on event order. Could be s2, s3, s2, or could be s2, s1, s2	p1, p2 or p2, p1
6	s2, s3	p1
7	s2, s1	p2, p4
8	do nothing	null, null
9	Cannot show. Depends on event order. Could be s3, s4, s3 could be s3, s2, s3	p1, p2 or p2, p1
10	s3, s4	p1
11	s3, s2	p2
12	do nothing	null, null
13	impossible	
14	impossible	
15	s4, s3	p2
16	do nothing	null, null

4.4 Deriving Test Cases from a Finite State Machine

Finite state machines offer excellent support for identifying test cases for event-driven systems; however, they are almost useless for computational applications such as the Insurance Premium Problem.

Definition: In a *system-level finite state machine* (SL/FSM), states can be either contexts of events or stages of processing, and transitions are caused by input events. Transitions optionally can show output events as the "action" part of the transition.

In a system-level finite state machine, paths from start to sink states are nearly identical to long use cases and to system level test cases. The start states and sink states correspond to the pre and postconditions of a use case. The event sequence is easily derived from the causes and actions of transitions. The only thing missing is some narrative description of the use case, and maybe a descriptive name. A system-level finite state machine directly supports the following test coverage metrics:

- Every state
- Every input event
- Every output event
- Every transition
- Every path (problematic if there are loops in the SL/FSM)

4.4.1 The Insurance Premium Problem

Inputs	Outputs (Actions)	States (Stages of Processing)
e1: base rate	a1: base rate × 1.5	s1: Idle
e2: 16 <= age < 25	a2: base rate × 1.0	s2: Apply age multiplier
e3: 25 <= age < 65	a3: base rate × 1.2	s3: Apply claims penalty
e4: age >= 65	a4. add $0	s4: Apply goodStudent reduction
e5: at fault claims = 0	a5. add $100	s5: Apply nonDrinker reduction
e6: 1 <= at fault claims <= 3	a6. add $300	s6: Done
e7: 4 <= at fault claims <= 10	a7, subtract $0	
e8: goodStudent = T	a8. subtract $50	
e9: goodStudent = F	a9. subtract $75	
e10: nonDrinker = T		
e11: nonDrinker= F		

As a finite state machine, the one in Figure 4.12 is almost useless for deriving test cases. It is technically correct, but the flowchart version is more helpful. There are 36 distinct paths through this finite state machine, and they correspond directly to the 36 paths in the flowchart (Figure 2.7) and to the 36 rules in the Mixed Entry Decision Table formulation in Chapter 3. (There are no paths for the cases in which insurance is denied.) As the finite state machine adds nothing to the earlier versions, it will not be used in the chapters on the more sophisticated transition-based models. In fact, finite state machines are a special case of a Petri net (Chapter 5).

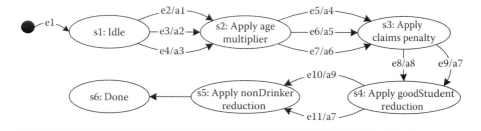

Figure 4.12   The Insurance Premium Problem finite state machine.

Table 4.9   Mapping Paths in a Finite State Machine to State Sequences

	Sample Garage Door Finite State Machine Paths (As State Sequences)	
Path	*Description*	*State Sequence*
1	Normal door close	s1, s5, s2
2	Normal door close with one intermediate stop	s1, s5, s3, s5, s2
3	Normal door open	s2, s6, s1
4	Normal door open with one intermediate stop	s2, s6, s4, s6, s1
5	Closing door with light beam interruption	s1, s5, s6, s1

The paths in Table 4.9 correspond directly to those derived from the decision table formulation in Table 3.15. Table 4.10 illustrates the detailed use case for a "soap opera" path (one that is as long as possible). The state sequence is <s1, s5, s3, s5, s6, s4, s6, s1>. This path covers all but one of the states and all but two of the input events (Figure 4.13).

4.4.2 The Garage Door Controller

Input Events	*Output Events (Actions)*	*States*
e1: control signal	a1: start drive motor down	s1: Door Up
e2: end of down track hit	a2: start drive motor up	s2: Door Down
e3: end of up track hit	a3: stop drive motor	s3: Door stopped going down
e4: laser beam crossed	a4: reverse motor down to up	s4: Door stopped going up
		s5: Door closing
		s6: Door opening

Table 4.10   Soap Opera Use Case for a Long Path in Figure 4.13

Use Case Name:	Garage Door Controller Soap Opera Use Case	
Use Case ID:	GDC-UC-1	
Description:	Use case for the state sequence <s1, s5, s3, s5, s6, s4, s6, s1>	
Preconditions:	1. Garage Door is Up	
Event Sequence:	**Input Event**	**System Response**
	1. e1: control signal	2. a1: start drive motor down
	3. e1: control signal	4. a2: stop drive motor
	5. e1: control signal	6. a1: start drive motor down
	7. e4: laser beam crossed	8. a4: reverse motor down to up
	9. e1: control signal	10. a2: stop drive motor
	11. e1: control signal	12. a2: start drive motor up
	13. e3: end of up track hit	14. a3: stop drive motor
Postconditions:	1. Garage Door is Up	

4.5 Lessons Learned from Experience

I was part of a team that proposed to model a small telephone switching system (a Private Automatic Branch Exchange, or PABX) with finite state machines. We ended up with more than 200 states: we called them "mega-states" because we described the internal behavior of a mega-state with decision tables. The states, transition-causing events, and both prescribed and proscribed behaviors were all defined textually—a drawing with so many states was deemed useless. A colleague with three decades of telephony experience was the technical advisor for hardware-related questions. He would occasionally taunt us "college boys" to do some real work. We usually responded by telling him that, once we were done, we would generate more test cases than he would ever think of. We built a program to generate all the paths through the finite state machine (in retrospect, this was a true MBT project). Each path included the inputs and outputs on the transitions, as well is the initial (pre-conditions) and terminal (post-conditions) states—a nearly complete system level test case. The program generated on the order of 3000 paths, which we dumped on our taunting colleague's desk. At first, he was both excited

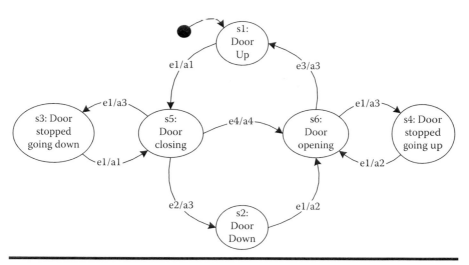

Figure 4.13   The Garage Door Controller finite state machine.

and enthusiastic, sometimes rushing in with what he felt was a really interesting test case. Later, the enthusiasm stopped. One day our colleague came in telling us to scrap the whole thing—one of the generated paths was technically impossible. On further analysis, it turned out that there was a deep, subtle hardware dependence between a pair of states. We responded by saying that we would run the path generation program again, this time deleting any path that traversed the pair of dependent states. We had no answer when our colleague asked about the possibility of additional pairs of dependent states. This was our first encounter with what we later named the "Sorcerer's Apprentice Problem" [Jorgensen 1985]. The name came from the Walt Disney cartoon in which the Sorcerer's Apprentice has the broom to carry pails of water, but the Sorcerer's Apprentice cannot control the consequences, and sets in motion a process much larger than planned. There is a form of leverage when using finite state machines—defining that a small number of states and transitions can generate a surprisingly large number of paths. The problem my group had was that we were on the wrong end of the modeling lever, with 3000+ paths to check out. There may have been one additional pair of dependent states, but on the whole, the experiment was successful. It was certainly a learning experience.

Recently I had a university-based experience. I asked my undergraduate class to provide a set of Use Cases for the Garage Door Controller. Several students submitted use cases for the "stopped while closing" and "stopped while opening" cases in which their postcondition was "Door stopped part-way." The next assignment asked them to derive finite state machines from their use cases, and then to merge them into a full finite state machine. Those with the "stopped part-way" postconditions ended up with the partial finite state machines as in Figure 4.14. When the finite state machines were combined, the result, in Figure 4.15, was nondeterministic, as in Figure 4.5. What happens when e1 occurs in state s3? This illustrates, in a small way, the greater problem of developing finite state machines in a "bottom-up" way—a process that characterizes the agile methods.

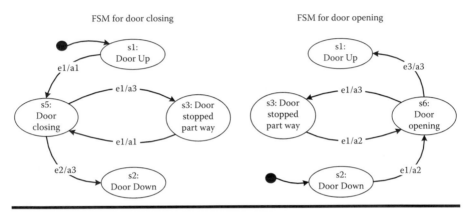

Figure 4.14 Separately developed finite state machines.

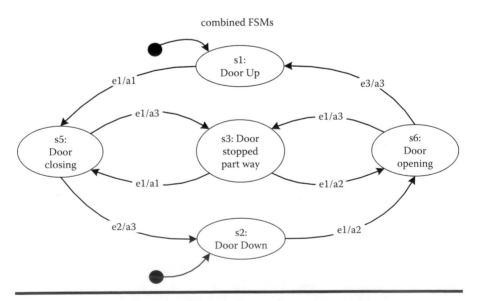

Figure 4.15 Combining separately developed finite state machines.

4.6 Advantages and Limitations

Finite state machines are intuitively understandable by both developers and customers. When displayed in graphical form, paths are easily seen, and because this is a two-dimensional representation, related paths (test cases) can be identified visually. As finite state machines can express both prescribed and proscribed behaviors, they are especially amenable to analysis by execution engines. Also, finite state machines can be extended by associating probabilities with transitions. Doing so yields a notion of execution time traffic behavior, which, in turn, supports the notion of operational profiles in system testing. Finally, if cost penalties can be associated with paths, the costs and probabilities will support risk-based testing.

On the downside, finite state machines are inherently vulnerable to the famous state machine explosion. We saw an example of this in Figure 4.6. Figure 4.9 shows the effect of considering both expected and unexpected events in every state. In Chapter 7 (Statecharts), we will see that David Harel's notation deals very effectively with both the state explosion and the spaghetti-like tangle of edges when every event is permitted in every state.

Finite state machines are just one of several transition-based models. The others covered here are Petri nets (Chapter 5), Event-Driven Petri Nets (Chapter 6), their extension to Swim Lane Event-Driven Petri Nets (Chapter 8), and Statecharts (Chapter 7). Figure 4.16 shows a lattice of these transition-based models, where edges refer to the originating model being "more expressive than" the model on the terminating end of an edge. Table 4.11 from

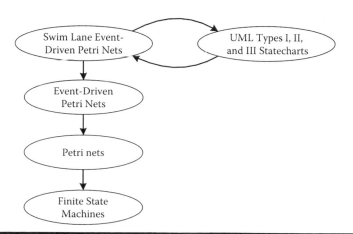

Figure 4.16   Lattice of expressive power of transition-based models.

Table 4.11   Representation of Behavioral Issues with Finite State Machines

Issue	Represented?	Example
Sequence	Yes	
Selection	Awkward	Must use alternative events to cause distinct transitions
Repetition	Yes	
Enable	No	The attendant or credit card approval, also the trigger squeeze event enables the display process
Disable	No	The 4% level event. Also the trigger release event disables the display process
Trigger	No	The trigger squeeze event triggers the flow of fuel
Activate	No	The enable and disable primitives combine to be an activate
Suspend	No	
Resume	No	
Pause	No	
Conflict	No	
Priority	No	
Mutual exclusion	No	
Concurrent execution	No	Must allocate finite state machines to separate devices
Deadlock	No	Observed in an execution table
Context-sensitive input events	Yes	The coin insertion events in the espresso vending machine
Multiple-context output events	Yes	The dispense espresso event in the espresso vending machine
Asynchronous events	No	
Event quiescence	No	

[Jorgensen 2009] shows the nineteen selected behavioral issues, and which of these can be expressed with finite state machines. This table will be repeated for each of the transition-based models in the following chapters.

References

[Brooks 1986]
Brooks, Jr., Frederick P., No Silver Bullet - Essence and Accident in Software Engineering, *Proceedings of the IFIP Tenth World Computing Conference*, H.-J. Kugler, ed., Elsevier Science B.V., Amsterdam, the Netherlands, 1986, pp. 1069–1076. [also in The Mythical Man-Month]
[Jorgensen 1985]
Jorgensen, Paul C., Complete Specifications and the Sorcerer's Apprentice Problem, *Proceedings of the International Computers and Applications Conference*, (*COMPSAC'86*), Chicago, IL, October 1986.
[Jorgensen 2009]
Jorgensen, Paul C., *Modeling Software Behavior—A Craftsman's Approach.* CRC Press, Boca Raton, FL, 2009.

Chapter 5

Petri Nets

Petri nets were the topic of Carl Adam Petri's PhD dissertation in 1963; today they are the accepted model for protocols and applications involving decision-based processing. Although the diagrams do not scale up well, there is an elegant database solution to this discussed toward the end of this chapter. We saw that finite state machines easily lead to a state explosion, particularly when two communicating finite state machines are merged via a cross product. This is accomplished in a more elegant way with Petri nets. (Much of Sections 5.1 and 5.2 are adapted from *Modeling Software Behavior—A Craftsman's Approach* [Jorgensen 2009].)

5.1 Definition and Notation

Petri nets are a special form of directed graph: a bipartite directed graph. A bipartite directed graph has two sets of nodes, V_1 and V_2, and a set of edges E, with the restriction that every edge has its initial node in one of the sets V_1, V_2, and its terminal node in the other set. In a Petri net, one of the sets is referred to as "places," and the other is referred to as "transitions". These sets are usually denoted as P and T, respectively. Places are inputs to and outputs of transitions; the input and output relationships are functions, and they are usually denoted as In and Out, as in the following definition.

Definition: A *Petri net* is a bipartite directed graph (P, T, In, Out), in which P and T are disjoint sets of nodes, and In and Out are sets of edges, where In $\subseteq$ P $\times$ T, and Out $\subseteq$ T $\times$ P.

For the sample Petri net in Figure 5.1, the sets P, T, In, and Out are

P = {p1, p2, p3, p4, p5}
T = {t1, t2, t3}

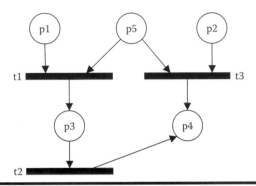

Figure 5.1   A Petri net.

In = {<p1, t1>, <p5, t1>, <p5, t3>, <p2, t3>, <p3, t2>}
Out = {<t1, p3>, <t2, p4>, <t3, p4>}

In this chapter, the names given to Petri net places and transitions are arbitrary. We frequently use s and t for transitions, but other (possibly more mnemonic) names are acceptable; similar comments apply to place names. Places are usually drawn as circles and transitions as rectangular bars. In Europe, transitions are frequently open rectangles. Petri nets are executable, in more interesting ways than finite state machines. The next few definitions lead us to Petri net execution.

Definition: A *marked Petri net* is a five-tuple (P, T, In, Out, M) in which (P, T, In, Out) is a Petri net and M is a set of mappings of places to positive integers.

The set M is called the marking set of the Petri net. Elements of M are n-tuples, where n is the number of places in the set P. For the Petri net in Figure 5.1, the set M contains elements of the form <n1, n2, n3, n4, n5>, where the n's are the integers associated with the respective places. The number associated with a place refers to the number of tokens that are said to be "in" the place. Tokens are abstractions that can be interpreted in a variety of modeling situations. For example, tokens might refer to the number of times a place has been used, or the number of things in a place, or whether the place is true. Figure 5.2 shows a marked Petri net.

The marking tuple for the marked Petri net in Figure 5.2 is <1, 1, 0, 2, 0>. We need the concept of tokens to make two essential definitions.

5.1.1 Transition Enabling and Firing

Definition: A transition in a Petri net is *enabled* if there is at least one token in each of its input places.

There are no enabled transitions in the marked Petri net in Figure 5.2. If we put a token in place p3, then transition t2 would be enabled.

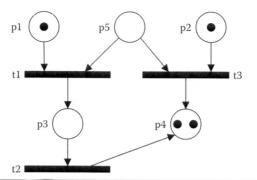

Figure 5.2   A marked Petri net.

Definition: When an enabled Petri net *transition fires*, one token is removed from each of its input places, and one token is added to each of its output places.

In Figure 5.3, transition t2 is enabled in the left net, and it has been fired in the right net. The marking set for the net in Figure 5.3 contains two tuples, the first shows the net when t2 is enabled, and the second shows the net after t2 has fired.

$$M = \{<1, 1, 1, 2, 0>, <1, 1, 0, 3, 0>\}$$

Tokens may be created or destroyed by transition firings. Under special conditions, the total number of tokens in a net never changes; such nets are called conservative. We usually will not worry about token conservation. Markings let us execute Petri nets in much the same way that we execute finite state machines. (In fact, finite state machines are a special case of Petri nets.) Some formulations of Petri nets associate a weight with each input edge to a transition. The weight, a natural number, signifies the number of tokens that must be in the input place in order for that place to contribute to enabling. Similarly, there can be a weight on an output edge that signifies

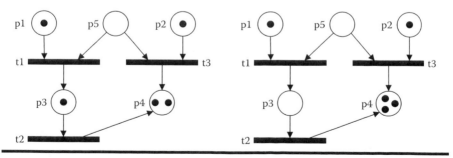

Figure 5.3   Before (left) and after (right) firing t2.

how many tokens are placed in an output place when the transition fires. We will not make use of these extensions.

5.1.2 Conventions

Convention 1: Regardless of how many transitions are enabled, only one transition at a time can be fired.

Many authors assert that Petri nets are the correct representation of concurrency, but as simultaneous transition firing is not allowed, there can be no true concurrency. Consider the difference between an unmarked Petri net and any of its marking sequences. This distinction is isomorphic to the distinction between the intension of a database model, and the many possible extensions that result from different population sequences. Parallel paths in an unmarked Petri net show potential concurrency, but the single transition firing restriction precludes concurrent execution. As an unmarked Petri net can have many possible marking sequences, these represent possible executions, but still not concurrent executions.

Convention 2: Places may be inputs to more than one transition; symmetrically, places may be outputs of more than one transition.

This multiplicity is much of what makes Petri nets so useful (and more powerful than finite state machines).

Convention 3: Transitions must have at least one input place and at least one output place; they may have more than one input place, and more than one output place.

Some formulations show transitions with no input places. Apparently, these transitions are always enabled. A transition with no output places is curious—what happens if it is fired? How could we ever know that such a transition has been fired?

5.1.3 Nongraphical Representations

5.1.3.1 Textual Representation

Figure 5.2 can also be represented in a template, as shown below.

```
Transition <transition Id>    <transition name>
      has input places (repeat for each input place):
            <place 1>    <place name> with marking    <number of tokens>
      and has output places (repeat for each output place):
            <place 1>    <place name> with marking    <number of tokens>
```

Here is the template completed for Figure 5.2:

Transition <t1> <(not given)>
    has input places (repeat for each input place):
        <p1> <(not given)> with marking < 1 >
        <p5> <(not given)> with marking < 0 >
    and has output places (repeat for each output place):
        <p3> <(not given)> with marking < 0 >

Transition <t2> <(not given)>
    has input places (repeat for each input place):
        <p3> <(not given)> with marking < 0 >
    and has output places (repeat for each output place):
        <p4> <(not given)> with marking < 2 >

Transition <t3> <(not given)>
    has input places (repeat for each input place):
        <p5> <(not given)> with marking < 0 >
        <p2> <(not given)> with marking < 1 >
    and has output places (repeat for each output place):
        <p4> <(not given)> with marking < 2 >

5.1.3.2 Database Representation

The biggest problem with Petri nets is that the graphical version does not scale up well. This is/can be mitigated by using a database with relations for the information in a Petri net, and then developing sets of useful queries. An Entity–relationship (E/R) model of such a database is given in Figure 5.4. The database possibility completely resolves the scale-up problem but the resolution is at the expense of visual clarity. Closely related to this, most users find that they use a key (or legend) to give short names to Petri net places and transitions, thereby improving communication.

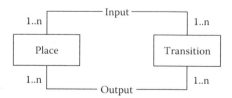

Figure 5.4   E/R model of a Petri net database.

The entities and relations for the Petri net in Figure 5.3 (in table form) are

Place		
Place	Name	Tokens
p1	(not given)	1
p2	(not given)	1
p3	(not given)	0
p4	(not given)	2
p5	(not given)	0

Transition		
Transition	Name	Enabled?
t1	(not given)	No
t2	(not given)	No
t3	(not given)	No

Input	
Transition	Place
t1	p1
t1	p5
t2	p3
t3	p2

Output	
Transition	Place
t1	p3
t2	p4
t3	p4

The information in this populated E/R model is both necessary and sufficient to recreate a Petri net equivalent (with the exception of spatial placement) to that in Figure 5.2. It seems excessive for this simple example, but this resolves the scale-up problem of larger Petri nets.

5.2 Technique

Because Petri nets are less well understood than finite state machines, we will look extensively at questions of technique. For starters, here are some general hints:

1. Use transitions to represent actions that have inputs and outputs.
2. Use places to represent any of the following:
   - Data
   - Pre- and postconditions
   - States
   - Messages
   - Events
3. Use the input relationship to represent prerequisites and inputs.
4. Use the output relationship to represent consequences and outputs.
5. Use markings to represent "states" of a net, memory, or counters.

6. Tasks can be considered to be individual transitions or subnets.
7. Subsets of input places to a transition can be used to define a context for an input event.
8. Use Petri nets to focus on complex portions of the system under test (SUT).
9. Use marking sequences to generate test scenarios.

In the following subsections, we look at Petri net formulations of 14 of the 19 modeling issues identified in [Jorgensen 2009]. We could refer to these as "Petri patterns," along the lines of design patterns. They are listed here by categories. The ESML prompts are from the Extended Systems Modeling Language [Bruyn 1988]. They are also present in the early versions of the StateMate product for Statecharts. The event-related issues are not well represented by ordinary Petri nets, but they will be represented by Event-Driven Petri Nets in Chapter 6.

- Structured programming constructs
  - Sequence
  - Selection
  - Repetition
- ESML prompts
  - Enable
  - Disable
  - Activate
  - Trigger
  - Suspend
  - Resume
  - Pause
- Task interaction (Petri net primitives)
  - Conflict
  - Priority
  - Mutual exclusion
  - Synchronization (start, stop)
- Events
  - Context-sensitive input events
  - Multiple cause output events
  - Asynchronous events
  - Event quiescence

5.2.1 Sequence, Selection, and Repetition

Sequence and repetition are easily shown in a Petri net; see for example Figure 5.5. Transitions s1 and s2 are in sequence, and they are also in a loop. Selection is not as simple. Basically, there are two choices, as shown in Figure 5.6.

In the left Petri net, the actual selection is made by which place is marked, p1 or p2 (they are in an exclusive-or relation). The transitions s1 and s2 perform a simple

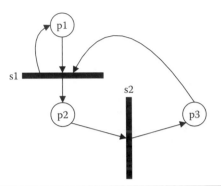

Figure 5.5   Sequence and repetition in a Petri net.

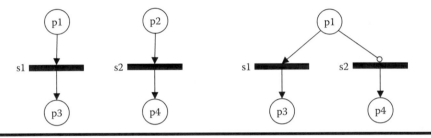

Figure 5.6   Two forms of Petri net selection.

calculation, and leave the results in places p3 and p4. The right net uses an "inhibitor arc" that is denoted by a little circle at the end rather than an arrowhead. Inhibitor arcs contribute to transition enabling in a negative way: if the place is not marked, the inhibitor arc helps enable the transition to which it is connected. So if p1 refers to some proposition p, marking place p1 is interpreted as the proposition p being true, and no marking refers to p being false. Which form of selection is better is really a personal and stylistic choice. I prefer the simpler alternative (as in the left part of Figure 5.6).

5.2.2 Enable, Disable, and Activate

Definition

> *Enable*: When activity A enables activity B, activity B may execute, but not necessarily immediately.
>
> *Disable*: When activity A disables activity B, activity B may not execute. The termination is immediate, which makes disable look like an "untrigger" (defined soon).
>
> *Activate*: An enable/disable sequence performed by activity A for/on activity B. Presumably there is a time interval in which B does what it has been enabled to do, and is then disabled.

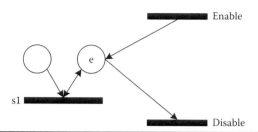

Figure 5.7 Petri nets for Enable, Disable, and Activate.

The task s1 in Figure 5.7 cannot fire unless it has been enabled by the enable task. Once s1 is enabled, it reenables itself every time it fires. The disable task is in Petri net conflict with s1 with respect to the enable place. If the disable transition fires, s1 is disabled.

5.2.3 Trigger

Definition
> *Trigger*: When activity A triggers activity B, B immediately executes.

Task s1 in Figure 5.8 is partially enabled. When the trigger transition fires, the trigger place is marked, then s1 can fire. There is no convention regarding how long a transition "remains" triggered. The two-headed arrow in Figure 5.8 lets the transition s1 continue to fire until the other inputs are no longer available. If the connecting arrow were just the usual one, s1 could fire only once.

5.2.4 Suspend, Resume, and Pause

Definition
> *Suspend*: When activity A suspends activity B, activity B no longer executes, but intermediate work is saved.
> *Resume*: When activity A resumes activity B, activity B (which has been suspended) starts with the partially completed work that was in progress when it was suspended.
> *Pause*: A suspend/resume sequence performed by activity A for/on activity B. Presumably there is a time interval in which B does nothing.

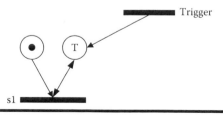

Figure 5.8 Petri net for the ESML Trigger.

The intent of the ESML Suspend and Resume prompts is that they interrupt an executing task, and then restart it "where it left off." Task s1 is subdivided into three (an arbitrary choice) subtasks, as shown in Figure 5.9 to illustrate the "where it left off" part of the problem. Notice that the suspend transition has an interlock (see Figure 5.10) with the resume transition, so the suspend transition must fire first. The suspend transition is in Petri net conflict with some intermediate task. Firing the suspend transition disables the intermediate subtask (s1.2 here), and firing the resume transition reenables s1.2.

The ESML Activate prompt is just an enable-disable sequence; as with Activate, the ESML Pause prompt is just a suspend-resume sequence.

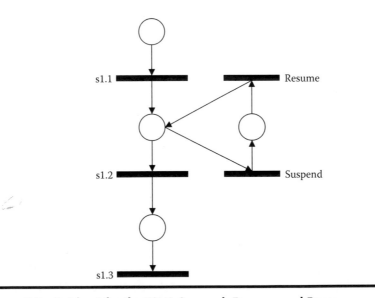

Figure 5.9   Petri net for the ESML Suspend, Resume, and Pause.

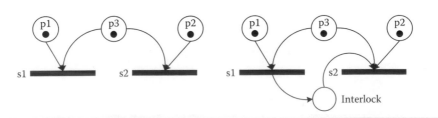

Figure 5.10   Petri net conflict and Interlock.

5.2.5 Conflict and Priority

Petri net conflict exhibits an interesting form of interaction between two transitions. Transitions s1 and s2 are both enabled in the left side of Figure 5.10. Firing either transition disables the other. This is the classic form of Petri net conflict. More specifically, we say that transitions s1 and s2 are in conflict with respect to place p3. One way to break a conflict is to use an interlock place that forces transition s2 to wait until transition s1 has fired, as in the right side of Figure 5.10.

5.2.6 Mutual Exclusion

In Figure 5.11, the sequence <t1.1, t1.2, t1.3> and the sequence <t2.1, t2.2, t2.3> are mutually exclusive. As t1.1 and t2.1 are in Petri net conflict with the semaphore place (S), firing one disables the other, and the exclusion is continued until the end of the excluded region (at either t1.3 or t2.3). Notice that the semaphore can be understood as a "bidirectional interlock."

5.2.7 Synchronization

Tasks 1, 2, and 3 are given a simultaneous start by the "start" task in Figure 5.12. To be truly parallel, the tasks would have to be in separate devices; this is not shown in the Petri net. This is sometimes described as a synchronized start.

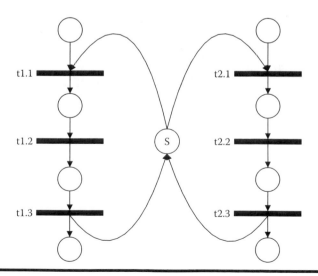

Figure 5.11   Petri net for mutual exclusion.

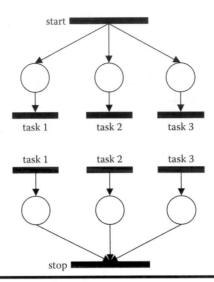

Figure 5.12   Synchronized start and stop.

In the lower part of Figure 5.12, the "stop" task cannot execute until each of its prerequisite tasks (tasks 1, 2, and 3) is complete. The Two-Phase Commit protocol for updates in a distributed database is exactly represented by the synchronized starts and stops as shown in Figure 5.12.

5.2.8 Some Consequences of Marking and Enabling

Marking places in a Petri net is a graphical/numerical way to represent various possible execution sequences of an unmarked Petri net. In Figure 5.13, we revisit the producer portion of the producer–consumer problem in Figure 5.15.

As place p1 is marked (left net), transition t1 is enabled. If we fire t1, place p1 is no longer marked, but place p2 is marked (center net). Now transition t2 is enabled.

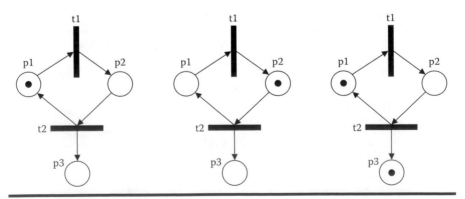

Figure 5.13   The Producer Petri net.

Table 5.1 A Sample Execution Table of the Producer Petri Net

Execution Step	Marking <p1, p2, p3>	Enabled Transitions	Fired Transition
m0	<1, 0, 0>	t1	t1
m1	<0, 1, 0>	t2	t2
m2	<1, 0, 1>	t1	t1
m3	<0, 1, 1>	t2	t2
m4	<1, 0, 2>	t1	t1
m5	<0, 1, 2>	t2	t2

If we fire t2, place p2 is no longer marked, but places p1 and p3 are marked. We could continue firing t1 and t2, and after each cycle, p3 would have an additional marking token. Table 5.1 shows a few time steps of this.

First observation: Either t1 or t2 is always enabled, so we could execute this Petri net forever.

Definition: A *live* Petri net is one in which some transition is always enabled.

Definition: In a Petri net, *deadlock* occurs when no transition is enabled.

Definition: A Petri net is *conservative* if the sum of all tokens in the net is constant.

Liveness depends on the initial marking. In our example, if the initial marking was just p3, no transition would be enabled. If we removed place p3 from the Petri net in Figure 5.13, there would always be exactly one token in the live.

5.2.9 Petri Nets and Finite State Machines

Finite state machines are a special case of Petri nets in which every Petri net transition has exactly one input place and exactly one output place (see Figure 5.14). To make a Petri net out of a finite state machine, consider the states to be places, and the transitions to be Petri net transitions. "Transitions" is an overloaded term: in a finite state machine they are caused by something, and can generate outputs. In the finite state machine notation, this is the "fraction" that annotates a state transition. In a Petri net, the states of a finite state machine become Petri net places, and the events and actions on finite state machine transitions become Petri net transitions, with the special property that the Petri net transitions have exactly one input place and one output place. States s1, s2, s3, and s4 from the finite state machine are mapped to places s1, s2, s3, and s4 in the Petri net. We can add places in the Petri net that serve as memory, in this case, the places labeled "from s2" and "from s3." The transitions from place s4 to places s5 and s6 now use the memory, so everything is correct.

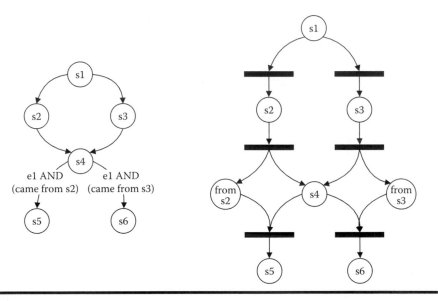

Figure 5.14   Converting a finite state machine to a Petri net, and showing memory.

The finite state machine in the left side of Figure 5.14 is copied from Figure 4.2; notice that Petri nets resolve the "no memory" problem of finite state machines.

5.2.10 Petri Net Engines

Petri net engines are similar to finite state machine engines, at least at the interactive level. Given a Petri net, its possible execution tables indicate how a Petri net engine will execute a Petri net. (See Table 5.2 for a short execution table for the producer–consumer problem.) A Petri net engine needs a definition of all places, and the input and output functions relating places to transitions, and an initial marking, like the one in Table 5.2. Then the engine (and the user) proceeds as follows:

1. The engine determines a list of all enabled transitions.
2.1. If no transition is enabled, the net is deadlocked.
2.2. If only one transition is enabled, it is automatically fired, and a new marking is generated.
2.3. If more than one transition is enabled, the user must select one to fire, and then a new marking is generated. (Note that the user resolves any Petri net conflict.)
3. The engine returns to step 1 and repeats until either the user stops or the net is deadlocked.

**Table 5.2   A Sample Execution Table of the
Producer–Consumer Petri Net**

Execution Step	Marking	Enabled Transitions	Fired Transition
m0	<0, 1, 0, 1, 0, 0, 1>	t2	t2
m1	<1, 0, 1, 1, 0, 0, 1>	t1, t3, t5	t5
m2	<1, 0, 0, 1, 0, 0, 1>	t1, t6	t1
m3	<0, 1, 0, 1, 0, 1, 0>	t2, t6	t2
m4	<1, 0, 1, 1, 0, 1, 0>	t1, t3, t6	t3
m5	<1, 0, 0, 0, 1, 1, 0>	t1, t6	

There are seven levels of Petri net execution. Incidentally, they correspond exactly to execution levels for Statecharts (see Chapter 7).

Level 1 (*Interactive*): The user provides an initial marking and then directs transition firing, as described earlier.

Level 2 (*Burst Mode*): If there is a chain of steps in which only one transition is enabled, the whole chain is fired.

Level 3 (*Predetermined*): An input script marks places and directs execution.

Level 4 (*Batch Mode*): A set of predetermined scripts is executed.

Level 5 (*Probabilistic*): Similar to the interactive mode, only conflicts are resolved using transition firing probabilities (e.g., at step 2.3).

Level 6 (*Traffic Mix*): A demographic set of batch scripts, executed in random order.

Level 7 (*Exhaustive*): For a system with no loops, execute all possible "threads." Exhaustive execution is more sensibly done with respect to the reachability tree of a Petri net. If a system has loops, the use of condensation graphs can produce a loop-free version. This is computationally intense, as a given initial marking may generate many possible threads, and the process should be repeated for many initial markings.

5.3 Examples

5.3.1 The Producer–Consumer Problem

The Producer–Consumer problem is a Petri net classic. In Figure 5.15, nodes p1, p2, and p3, and transitions t1 and t2 represent the producer. Most texts present this as contention for a limited resource. My favorite interpretation is Carol making Swedish pancakes for David K and Paul (one at a time, as we shall see).

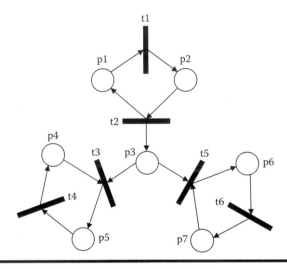

Figure 5.15   Producer–consumer Petri net.

The Producer (Carol) is represented by places p1, p2, and p3, and by transitions t1 and t2. Consumer 1 (David K) is represented by nodes p3, p4, and p5 and transitions t3 and t4. Similarly, Consumer 2 (Paul) is represented by nodes p3, p6, and p7 and transitions t5 and t6. Note the intense interaction of the people and roles at place p3.

Possible interpretations of the places and transitions are

p1: request for another pancake	t1: Carol pours pancake batter
p2: pancake in pan	t2: Carol fries and serves a pancake
p3: pancake on serving plate	t3: David K takes the pancake
p4: David K ready for another pancake	t4: David K eats pancake
p5: David K has a pancake	t5: Paul takes the pancake
p6: Paul has a pancake	t6: Paul eats the pancake
p7: Paul ready for another pancake	

With the initial marking in Figure 5.16, this is a "live" net: some transition will always be enabled. Carol will keep on making Swedish pancakes as long as David K and Paul keep eating them. In fact, even if they stop eating them, Carol can keep on making pancakes by repeatedly firing transitions t1 and t2. If she does, place p3 (the serving platter) will have many pancakes, and this is nicely shown by a marking sequence. Table 5.2 describes a Swedish Pancake scenario in which Carol

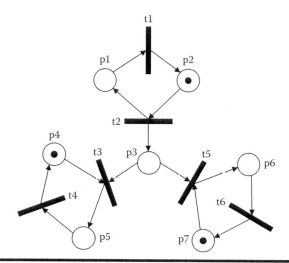

Figure 5.16   Possible initial marking for the producer–consumer Petri net.

makes a pancake (at m0), and David K and Paul are vying for the first pancake (at m1). Paul takes the pancake (firing t5 at step m1); Carol prepares to fry another pancake (at step m2). She fries and serves another pancake at step m3. At this point, only David K is enabled, and he takes the just-baked pancake (at step m4). In the last step (of this scenario) Carol is ready to make another pancake and Paul is still (politely) waiting.

This particular form of basic Petri nets is known as a "free choice" net, because when more than one transition is enabled, someone (the customer/ user!) chooses which transition to fire. There is extensive academic tool support for this form of Petri net execution, so this is clearly an executable specification. There are seven distinct levels of Petri net execution (see Section 5.5). Figure 5.17 is a slight change to show how marking sequences can be used. The edge from t2 to p1 is removed, and place p1 is marked with four tokens. This can be interpreted as batter sufficient for four Swedish pancakes. Now, regardless of David K and Paul consuming pancakes, Carol can only make four pancakes. After the fourth, transition t1 can never be enabled. The scenario can continue until David K and Paul finish the four pancakes, and then the net will be deadlocked.

5.3.2 *The Windshield Wiper Controller*

Table 5.3 lists the inputs and outputs, and the states for a Petri net model of the Windshield Wiper Controller. They are a repeat of the input events and actions in the finite state machine version of this problem in Chapter 4. The corresponding Petri net is shown in Figure 5.18. Look carefully at Figure 4.2 in Chapter 4—the

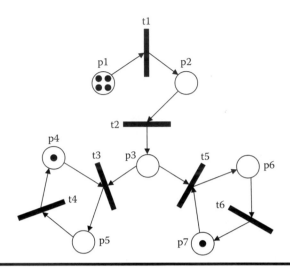

Figure 5.17   Initial marking for a net that will eventually be deadlocked.

Table 5.3   Windshield Wiper Controller Events and States

Inputs	Outputs	States
e1: move lever up one position	a1: deliver 0 wipes per minute	p1: Lever at OFF
e2: move lever down one position	a2: deliver 6 wipes per minute	p2: Lever at INT
e3: move dial up one position	a3: deliver 12 wipes per minute	p3: Lever at LOW
e4: move dial down one position	a4: deliver 20 wipes per minute	p4: Lever at HIGH
	a5: deliver 30 wipes per minute	p5: Dial at 1
	a6: deliver 60 wipes per minute	p6: Dial at 2
		p7: Dial at 3

states used in that figure, s1–s7, correspond exactly to the states p1 to p7 in Figure 5.18. Recall that, in Chapter 4, we went to great lengths to try to combine the Lever and Dial finite state machines, without much success. They are combined correctly in Figure 5.18, with the addition of transitions s11, s12, and s13 that deliver the three intermittent (INT) wiper speeds.

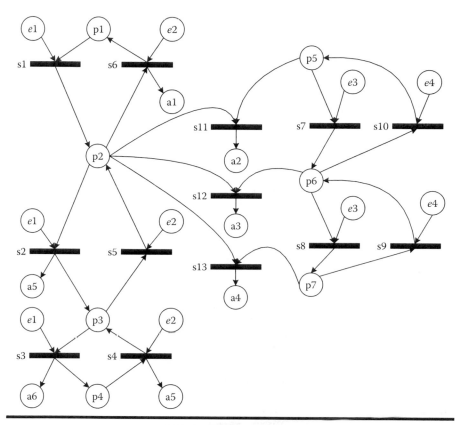

Figure 5.18   Windshield Wiper Controller Petri net.

There are several details that must be emphasized. The first is that it is a little artificial to show inputs several times—input e1 (move lever up one position) occurs three times, as does the corresponding input, e2. When the Petri net is drawn in this way, it is easier to follow an execution path, but it is harder to see all the contexts in which inputs can occur (See Figure 5.29 for another example of the single-occurrence style of input places.). The arrows connecting transitions s11, s12, and s13 to places p2, p5, p6, and p7 are double headed to allow them to remain marked when a dial input occurs and when one of s11, s12, or s13 fires. Finally, as the states p5, p6, and p7 are mutually exclusive, only one of the transitions s11, s12, and s13 can fire when the lever portion is in the Intermittent state (p2). Table 5.4 is a Petri net Marking Sequence that corresponds to Table 4.4 in Chapter 4. It is also an execution table.

There is yet another anomaly in Table 5.4. Beginning at step m5 of the marking sequence, outputs a2 and a5 are marked by step m17, outputs a2, a3, a4, a5, and a6 are marked. This is wrong, because the actions are/should be mutually exclusive. There is no way to remove tokens from a place that is never an input to a transition.

Table 5.4 Marking Sequence for the Petri Net in Figure 5.18

Marking Sequence	Places: <Inputs e1, e2, e3, e4; Outputs a1, a2, a3, a4, a5, a6; states p1, p2, p3, p4, p5, p6, p7>	Enabled Transitions	Fired Transition
m0	<Inputs (none); Outputs (none), states p1, p5>	(none)	(none)
m1	<Inputs e1; Outputs (none), states p1, p5>	s1	s1
m2	<Inputs (none); Outputs (none), states p2, p5>	s11	s11
m3	<Inputs (none); Outputs a2, states p2, p5>	(none)	(none)
m4	<Inputs e3; Outputs a2, states p2, p6>	s12	s12
m5	<Inputs (none); Outputs a2, a3, states p2, p6>	(none)	(none)
m6	<Inputs e1; Outputs a5, states p3, p6>	(none)	(none)
m7	<Inputs e1; Outputs a2, a3; states p2, p6>	s2, s11	s2
m8	<Inputs (none); Outputs a2, a3, a5; states p3, p6>	(none)	(none)
m9	<Inputs e1; Outputs a2, a3, a5; states p3, p6>	s3	s3
m10	<Inputs (none); Outputs a2, a3, a5, a6; states p4, p6>	(none)	(none)
m11	<Inputs e2; Outputs a2, a3, a5, a6; states p4, p6>	s4	s4
m12	<Inputs (none); Outputs a2, a3, a5, a6; states p3, p6>	(none)	(none)
m13	<Inputs e3; Outputs a2, a3, a5, a6; states p3, p6>	s8	s8
m14	<Inputs (none); Outputs a2, a3, a5, a6; states p3, p7>	(none)	(none)
m15	<Inputs e2; Outputs a2, a3, a5, a6; states p3, p7>	s5	s5

(Continued)

Table 5.4 (*Continued*) Marking Sequence for the Petri Net in Figure 5.18

Marking Sequence	Places: <Inputs e1, e2, e3, e4; Outputs a1, a2, a3, a4, a5, a6; states p1, p2, p3, p4, p5, p6, p7>	Enabled Transitions	Fired Transition
m16	<Inputs (none); Outputs a2, a3, a5, a6; states p2, p7>	s13	s13
m17	<Inputs (none); Outputs a2, a3, a4, a5, a6; states p2, p7>	s13	s13
m18	<Inputs e?; Outputs a2, a3, a4, a5, a6; states p2, p7>	s6, s13	s6
m19	<Inputs (none); Outputs a2, a3, a4, a5, a6; states p1, p7>	(none)	(none)

5.4 Deriving Test Cases from a Petri Net

Since finite state machines are a special case of (ordinary) Petri nets, all the observations of Section 4.4 apply here as well. But even ordinary Petri nets offer more support for deriving test cases.

- As with SL/FSMs, paths in a Petri net are candidate test cases.
- The flexibility of markings and input and output arrows gives Petri nets a form of memory, which is impossible in any finite state machine. .
- The Petri net primitives express many interesting testing situations (conflict, priority, mutual, and exclusion).
- Marking sequences can represent less obvious situations such as deadlock and livelock.
- Petri nets are easily composed (as in the Windshield Wiper example) so Petri nets that represent class level behavior can be composed to support testing at higher levels (e.g., interacting classes).

The biggest limitation of ordinary Petri nets is that they don't explicitly represent events (in fact, this is what led to the definition of Event-Driven Petri Nets). The most a place can represent is that an event has occurred, and this is when the "event place" is marked. A second limitation is that the difference between places that represent events, and those that represent states must be considered to recognize context-sensitive input events. Another problem is that, once an output place is marked, it remains marked, which is both misleading and incorrect. Similarly, if an input event occurs more than once, it must be repeated in the Petri net (e.g., see the treatment of event e1 in Figure 5.18).

Table 5.5   Use Case Derived from Figure 5.18

Use Case Name:	Windshield Wiper Controller Execution Table Example	
Use Case ID:	WWUC-1	
Description:	Operator moves through a typical sequence of lever and dial events. (w.p.m. is wiper strokes per minute)	
Preconditions:	1. Lever position is OFF (p1), Dial position is 1 (p5).	
Event Sequence:	**Input Event**	**System Response**
	1. e1: move lever up one position (p2)	2. s11: deliver 6 w.p.m. (a2)
	3. e3: move dial up one position (p6)	4. s12: deliver 12 w.p.m. (a3)
	5. e1: move lever up one position (p3)	6. s3: deliver 30 w.p.m. (a5)
	7. e1: move lever up one position (p4)	8. s3: deliver 60 w.p.m. (a6)
	9. e2: move lever down one position (p3)	10. s2: deliver 30 w.p.m. (a5)
	11. e3: move dial up one position (p7)	(no effect on wiper speed)
	12. e2: move lever down one position (p2)	13. s13: deliver 20 w.p.m. (a4)
	14. e2: move lever down one position (p1)	15. s6: deliver 0 w.p.m. (a1)
Postconditions:	1. Lever position is OFF (p1) Dial position is 3	

As with finite state machines, Petri nets lead directly to test cases. Recall the Use Case and Execution Table (Table 4.4) in Chapter 4. Table 5.5 is the corresponding use case derived from a path in Figure 5.18.

5.4.1 The Insurance Premium Problem

Here are the Petri net places and transitions needed to express the Insurance Premium Problem (see Figure 5.19).

Places		Transitions (Actions)
p1: base rate	p8: 4 <= at fault claims <= 10	t1: base rate × 1.5
p2: 16 <= age < 25	p9: age and claims adjusted rate	t2: base rate × 1.0
p3: 25 <= age < 65	p10: goodStudent = T	t3: base rate × 1.2
p4: age >= 65	p11: goodStudent = F	t4. add $0
p5: age adjusted rate	p12: nonDrinker = T	t5. add $100
p6: at fault claims = 0	p13: nonDrinker = F	t6. add $300
p7: 1 <= at fault claims <=3	p14: rate accumulation	t7. subtract $50
		t8. subtract $0
		t9. subtract $0
		t10. subtract $75

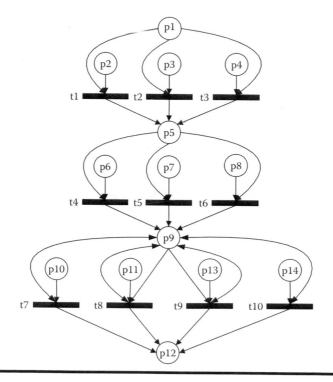

Figure 5.19　Insurance Premium Petri net.

Notice that the place pairs p10 and p11, and also p12 and p13, are needed to support the otherwise simple boolean decisions. The fact that the three age ranges, places p2, p3, and p4, are mutually exclusive is expressed by the Petri net conflict with respect to p1 of transitions t1, t2, and t3. A similar comment applies to the claims portion of the calculation. There is a subtlety about place p9 with respect to transitions t7, t8, t9, and t10–p9: it is both an input to and an output of the transitions. This is needed to support marking sequences that add both premium reductions, and the fact that they could be applied in either order. (The order is mandated in the flowchart description.) The Petri net formulation can also support the data flow view of testing. If there is a problem with place p2, for example, it will affect the rest of the calculation. Another example: suppose the final calculation is faulty. Then following the marking sequence that led to the faulty result in reverse order helps determine where the source of the fault is located. This is easy to do manually, but it would take sophisticated MBT tools to do such an analysis automatically.

The two Petri nets in Figure 5.20 show the initial marking sequences for two sample premium calculations—the least and the most expensive. David Harel makes a distinction between transformational and reactive programs [Harel 1988]. The Insurance Premium Problem is an example of a transformational program (all inputs are available before the calculation begins.) The initial marking sequences in Figure 5.20 show this, and also the order of the transitions preserves the correct calculation of order possibilities. The initial marking for the least expensive policy (left Petri net) results in the two possible transition firing

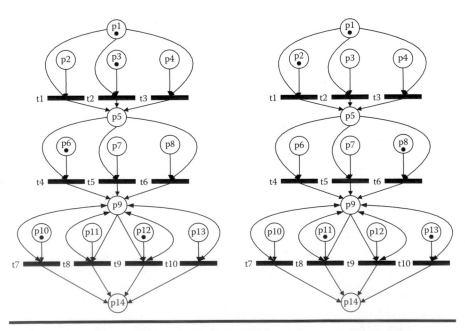

Figure 5.20 Initial markings for least (a) and most (b) expensive policies.

Table 5.6 Execution Table for Figure 5.20 (Left Side)

Execution Step	Marking <p1, p2, p3, p4, p5, p6, p7,p8, p9, p10, p11, p12, p13, p14, p15, p16>	Enabled Transitions	Fired Transition	Calculation Action
m0	p1, p3, p6, p10, p12	t2	t2	base rate × 1.0
m1	p5, p6, p10, p12	t4	t4	add $0
m2	p9, p10, p12	t7, t7	t7	subtract $50
m3	p9, p12, p14	t9	t9	subtract $75
m4	p14 (marked with 2 tokens)	t10	t10	accumulated premium

sequences: <t2, t4, t7, t9> or <t2, t4, t9, t7> depending on which discount is applied first. The first sequence refers to the case where p3 is true (25 ≤ age < 65), so transition t2 occurs (base rate × 1.0). Next, the goodStudent deduction (p10) so t7 occurs (subtract $50), followed by the nonDrinker reduction (p12) and transition t9 (subtract $75). The possible reductions are applied to the value that exists at p9; the final total is independent on the order in which reductions may be applied. That is why the seemingly duplicate transitions t8 and t9 (subtract $0) are present in the net (Table 5.6).

5.4.2 The Garage Door Controller

The input events, output events, and states from the finite state machine version all must be represented as places in an ordinary Petri net. This will clearly impact MBT test case derivation.

Input Events	Output Events	States
p1: Control signal	p5: Start drive motor down	p11: Door up
p2: End of down track hit	p6: Start drive motor up	p12: Door down
p3: End of up track hit	p7: Stop drive motor	p13: Door stopped going down
p4: Laser beam crossed	p8: Reverse motor down to up	p14: Door stopped going up
		p15: Door closing
		p16: Door opening

Transitions

t1: Door up to door closing	t6: Door down to door opening
t2: Door closing to door stopped going down	t7: Door opening to door stopped going up
t3: Door closing to door down	t8: Door stopped going up to door opening
t4: Door closing to door opening (safety reversal)	t9: Door opening to door up
t5: Door stopped going down to door closing	

There are several problems with the ordinary Petri net formulation:

1. Repeated control device signal places (p1) would all need to be marked.
2. The usual enabling and firing conventions result in persistently marked output places.
3. Marking sequences are/will always be awkward. The user/analyst will need to keep track of which places correspond to events that happened, and which are data/state oriented (see Figure 5.21).
4. Events not clearly shown.
5. Event quiescence is not easily recognized.

We can "fix" the problem of persistent tokens in output places (a work-around, actually) by adopting a convention is that, as the states of the drive motor are mutually exclusive, once an output event occurs, the corresponding Petri net transition will fire, unmarking the "present" output place and marking the "next" output place. A second work-around convention is to initially mark all instances of every

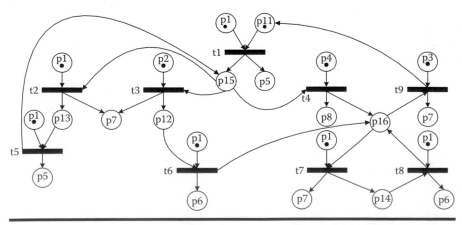

Figure 5.21 **Initial marking to show all potential input events and one initially marked *state* place.**

input event, and unmark them as individual input places that are used by transition firing. Figure 5.21 shows a possible way to deal with places that represent input events—they are all marked, plus an initial marking to begin Petri net execution (see Table 5.5) of the "soap opera" use/test case from Chapter 4.

In the initial marking M0, t1 is enabled, and the garage door is up.

Figures 5.22 through 5.28 show the marking sequence in Table 5.7.

Initial marking M0, t1 is enabled, door is up

Marking step M1: t1 has been fired. Transitions t2, t3, and t4 are enabled. Door is closing, p5 is marked (Figure 5.22).

Marking step M2: t2 fired. Transitions t3 and t4 are disabled. Output p5 is unmarked because output p7 is marked. Door is stopped going down. Transition t5 is the only enabled transition (Figure 5.23).

Marking step M3: t5 fired. Transitions t3 and t4 are enabled. Output p5 is marked because output p7 is unmarked. Door is going down (Figure 5.24).

Marking step M4: t4 fired. Transitions t7 and t9 are enabled. Output p8 is marked because output p5 is unmarked. Door is opening (Figure 5.25).

Marking step M5: t7 fired. Transition t8 is enabled; t9 is disabled. Output p7 is marked because output p8 is unmarked. Door is stopped going up (Figure 5.26).

Marking step M6: t8 fired. Transition t9 is enabled. Output p6 is marked because output p7 is unmarked. Door is going up (Figure 5.27).

Marking step M7: t9 fired. No transition is enabled. Output p7 is marked because output p6 is unmarked. Door is up (Figure 5.28).

The paths in Table 5.8 correspond directly to those derived from the decision table formulation in Table 3.15. Table 5.9 illustrates the detailed use case for a "soap opera" path (one that is as long as possible). The state sequence is <s1, s5, s3, s5, s6, s4, s6, s1>. This path covers all but one of the states and all but two of the input events.

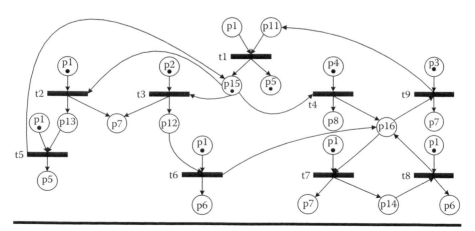

Figure 5.22   After firing t1, transitions t2, t3, and t4 are enabled.

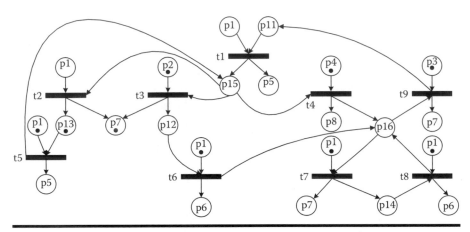

Figure 5.23   After firing t2, transition t5 is enabled, transitions t3 and t4 are disabled.

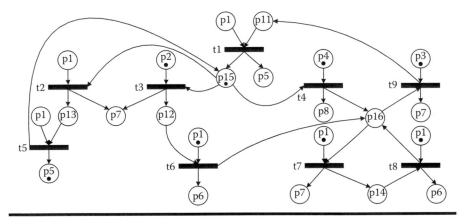

Figure 5.24   After firing t5, transitions t3 and t4 are enabled.

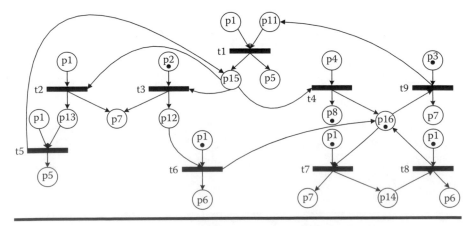

Figure 5.25   After firing t4, transitions t7 and t9 are enabled.

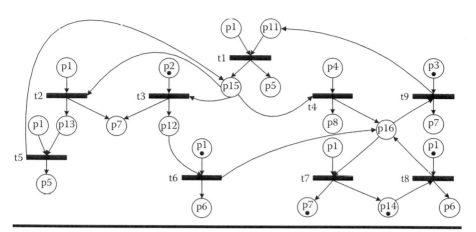

Figure 5.26   After firing t7, transition t8 is enabled.

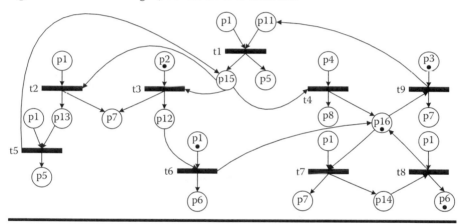

Figure 5.27   After firing t8, transition t9 is enabled.

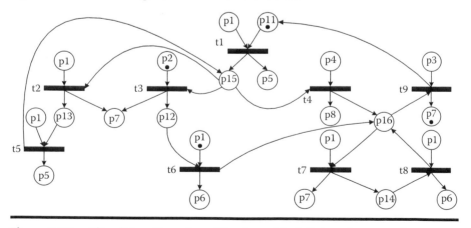

Figure 5.28   After firing t9, no transition is enabled. Point of event quiescence.

Table 5.7 Execution Table for Figures 5.22 through 5.28

Execution Step	Marking <Inputs p1, p2, p3, p4, Outputs p5, p6, p7, p8 states p11, p12, p13, p14, p15, p16>	Enabled Transitions	Fired Transition	Figure
m0	<p1(all instances), p2, p3, p4, Outputs (none), states p11>	t1	t1	5.23
m1	<p1(remaining), p2, p3, p4, Outputs p5, states p15>	t2, t3, t4	t2	5.24
m2	<p1(remaining), p2, p3, p4, Outputs p7, states p13>	t5	t5	5.25
m3	<p1(remaining), p2, p3, p4, Outputs p5, states p15>	t3, t4	t4	5.26
m4	<p1(remaining), p2, p3, Outputs p8, states p16>	t7, t9	t7	5.27
m5	<p1(remaining), p2, p3, Outputs p6, states p16>	t7	t8	5.29
m7	<p1(none), p2, Outputs p7, states p11>			5.30

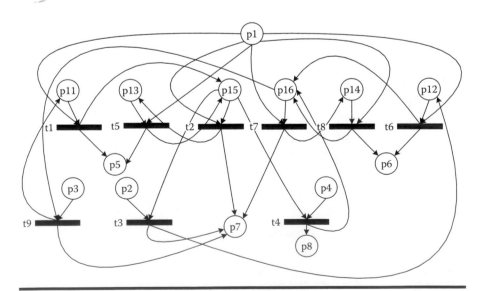

Figure 5.29 Garage Door Petri net without repeated input places.

Table 5.8   Sample Garage Door Controller Paths

Sample Garage Door Paths (as Transition Sequences)		
Path	*Description*	*Transition Sequence*
1	Normal door close	t1, t3
2	Normal door close with one intermediate stop	t1, t2, t5, t3
3	Normal door open	t6, t9
4	Normal door open with one intermediate stop	t6, t7, t8, t9
5	Closing door with light beam interruption	t1, t4, t9

Table 5.9   Soap Opera Use Case for the Long Path in Figures 5.21 through 5.28

Use Case Name:	Garage Door Controller Soap Opera Use Case (from a Petri net)	
Use Case ID:	GDC-UC-1	
Description:	Use Case for the Petri net Transition Sequence <t1, t2, t5, t4, t7, t8, t9>.	
Preconditions:	1. Garage Door is Up (p11)	
Event Sequence:	**Input Event (place)**	**Output Event (place)**
	1. p1: control signal	2. p5: start drive motor down
	3. p1: control signal	4. p7: stop drive motor
	5. p1: control signal	6. p5: start drive motor down
	7. p4: laser beam crossed	8. p8: reverse motor down to up
	9. p1: control signal	10. p7: stop drive motor
	11. p1: control signal	12. p6: start drive motor up
	13. p3: end of up track hit	14. p7: stop drive motor
Postconditions:	1. Garage Door is Up (p11)	

5.5 Lessons Learned from Experience

In one of the rewarding moments as a professor in the graduate program at my university, I received an e-mail from a former student who works at a food warehousing industry. He explained that he and some colleagues were wrestling with a complex problem, and just not being able to make much progress. He wrote:

> then I remembered those Petri nets we did in your class. It had been five years, but I dug out my notes, and modeled the problem. We had it solved in about 30 minutes.

I think the best use is for difficult, or maybe critical, situations, where the expressive power of Petri nets in both helpful, and needed.

Petri nets do not scale up well (Figure 5.30). They easily become so complex as to be useless. Here is a remarkable example of this [Kerth 2006]:

> *Together with our collaborating ecologists we develop a mechanistic model for the group decision making and fission-fusion pattern in a colony of Bechstein bats. Our collaborator provides recordings from a colony of tagged animals that allows [us] to trace position of each member of the colony over several weeks. In particular interest to us is to mathematically account for the decision*

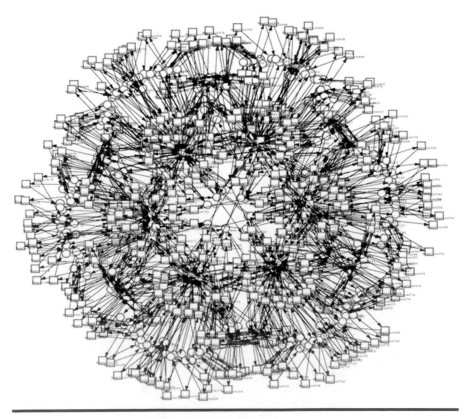

Figure 5.30 An exceedingly complex Petri net.

process involved in the determination of the new home roost of the colony. Our current model comprises established mechanism such as exploration, recruitment, quorum.

The modeling framework was chosen to be continuous and stochastic Petri nets. To illustrate the complexity of the model, below the Petri net for six roost locations and for a heterogenous population of two types is given.

5.6 Advantages and Limitations

Petri nets are much more popular in Europe than in the United States. They are extremely expressive, and they finally help us understand some of the nuances of the behavioral issues introduced in Chapter 3. Another advantage is that they constitute a bottom-up approach (appropriate for agile development), and can easily be composed into larger nets. As just noted, Petri net engines provide extensive analytic capabilities very early in a project, so they support model execution—a goal described in Chapter 1. Finally, they allow the recognition and analysis of situations such as starvation (conflict in which one transition always fires), deadlock, and mutual exclusion. Petri nets are probably best used for an extremely "close look" and important portions of a system.

Also, they do not deal with selection in a convenient way: outcomes of a decision must be input places to two transitions that are in conflict. Another disadvantage (resolved by Event-Driven Petri Nets) is that they do not represent event-driven systems well. In an execution table, or with an engine, there is no way to mark a place that corresponds to an input event. In that sense, they are "closed systems". The extension to Event-Driven Petri Nets (Chapter 6) resolves this, yielding "open systems." Also, once places that correspond to output events are marked, they remain marked. Finally, Petri nets are dismal at describing mathematical calculations. Technically, Petri nets do not represent concurrency—only one transition can be fired at a time. This can be circumvented by assigning subnets to separate devices. This deficiency has been elegantly resolved with Statecharts; we will see another resolution with Swim Lane Event-Driven Petri Nets in Chapter 8.

References

[Bruyn 1988]
Bruyn, W., R. Jensen, D. Keskar, and P. Ward, An extended systems modeling language based on the data flow diagram. *ACM Software Engineering Notes* 1988;13(1):58–67.
[Harel 1988]
Harel, David, On visual formalisms, *Communications of the ACM*, Vol. 31, No. 5, pp. 514–530, May, 1988.

[Jorgensen 2009]
Jorgensen, Paul C., *Modeling Software Behavior—A Craftsman's Approach*. CRC Press, Boca Raton, FL, 2009.
[Kerth]
Kerth, G., C. Ebert, and C. Schmidtke, Group decision making in fission-fusion societies: Evidence from two-field experiments in Bechstein's bats. *Proceedings of the Royal Society of London B: Biological Sciences* 2006;273(1602):2785–2790.

Chapter 6

Event-Driven Petri Nets

Inputs to any function can be either data, events, or some combination of these; similarly for outputs. Data is well understood, but events are not. There are two fundamental types of events—discrete and continuous. In terms of the examples in this book, the events in the Railroad Crossing Gate Controller, train arrivals and departures, and raising and lowering the crossing gate, are all discrete events. In the Windshield Wiper Controller, the input events (lever and dial changes) are discrete, and the output events (delivery of wiper strokes at different frequencies) are continuous. We usually think of input events as discrete, but an input representing something that is always being measured, even at different values, should be continuous. Examples? Fuel level in a tank, air speed, account balance, and so on. The question here is how models represent events.

For all their excellence at communication, flowcharts do a mixed job of representing events. The only way to identify the discrete/continuous dichotomy is via text. Further, events may appear as things that must happen (in a process box) or as outcomes of a decision. In flowcharts, events "just happen" and there is little additional information about them. The time order of events can be clearly represented, at least in terms of before and after. Decision tables are not much better. Events can be named, either in conditions (where they are input events) or in the action portion, where they are outputs. Since decision tables are (or should be) declarative, any sense of event sequence can be lost. This deficiency is offset by the ability to say that an input event either does or does not happen (the rule entries), and similarly for the output events in the action portion, where an output event either does or does not occur.

Finite state machines (FSMs) are a little better. We commonly say that state transitions are caused by events (making them input events) and output actions can be associated with transitions. Since paths through a finite state machine are directly related to the input/output sequence of a test case, at least we have the order of events. Composing independent finite state machines is awkward—the cross product of the states must be made, and this leads to the infamous "finite state explosion." There is still no distinction between discrete and continuous events. Ordinary Petri nets do not add much, partly because a finite state machine is a special case of a Petri net. The marking sequence that represents various possible executions of an ordinary Petri net is very attractive, but once a place corresponding to an output event is marked, it cannot be unmarked, thereby making it a continuous output event. This can be very awkward, particularly when output events are mutually exclusive, as they are with Windshield Wiper speeds. Also, the places in an ordinary Petri net have three interpretations—as data, as states, and as either input or output events. There is no visual distinction among these interpretations. Composition of ordinary Petri nets, however, is easy.

Event-Driven Petri Nets (EDPNs) were invented to explicitly clarify the representation of events. Both input and output events do not "just happen" as they do in finite state machines, decision tables, and flowcharts—they are explicitly shown. In Section 6.2, we will see the EDPN representations of the ESML prompts (begun in Chapter 5). In particular, we will be able to model why an event must (or must not) happen. There are separate symbols for input and output events, but there is still no discrete/continuous distinction, and this complicates marking sequences. Strictly speaking, both ordinary and Event-Driven Petri Nets can only show potential concurrency, because both have the convention that only one transition at a time can fire (execute). Separate Event-Driven Petri Nets can easily be composed, either visually in a drawing, or better, in a specially designed database. This is important, because there is no easy way to compose Statecharts, unless they have only one orthogonal region (and then they are at most, a hierarchic finite state machine). EDPNs have been extended to "swim lane Event-Driven Petri Nets", [DeVries 2013], and these have been proved to be equivalent to UML Types I, II, and III Statecharts. Composing Swim Lane EDPNs can be done in a database, but this is awkward in a drawing.

Both Statecharts and Swim Lane EDPNs have the ability to represent concurrent devices. The orthogonal regions of a Statechart and the "swim lanes" both represent separate, truly concurrent, devices. These models can show why an event happens, not just that it does happen. Statecharts, like finite state machines, have no specific symbol for events, so they must be recognized either by textual meaning or by some supplemental information to a Statechart diagram.

Basic Petri nets need two slight enhancements to become Event-Driven Petri Nets (EDPNs). The first enables them to express more closely event-driven systems, and the second deals with Petri net markings that express event quiescence, an important notion in object-oriented applications. Taken together, these extensions result in an effective, operational modeling view of software requirements for event-driven systems.

6.1 Definition and Notation

Definition: An *Event-Driven Petri Net (EDPN)* is a tripartite directed graph (P, D, T, In, Out) composed of three sets of nodes, P, D, and T, and two mappings, In and Out, where

- P is a set of port events
- D is a set of data places
- T is a set of transitions
- In is a set of ordered pairs from (P ∪ D) × T
- Out is a set of ordered pairs from T × (P ∪ D)

For the EDPN in Figure 6.2, the elements in the tuple (P, D, T, In, Out) are

P = {p1, p2, p3, p4}
D = {d1, d2, d3}
T = {s1, s2, s3, s4}
In = {<p1, s1>, <p1, s2>, <p2, s3>, <p2, s4>, <d1, s1>, <d2, s2>, <d3, s3>, <d2, s4>}
Out = {<s1, p3>, <s2, p4>, <s3, p3>, <s1, d2>, <s2, d3>, <s3, d2>, <s4, d1>}

EDPNs express four of the five basic system constructs defined in Chapter 1 (and repeated in Section 6.3); only devices are missing. The set T of transitions corresponds to ordinary Petri net transitions, which are interpreted as actions. There are two kinds of places, port events, and data places, and these are inputs to or outputs of transitions in T as defined by the input and output functions In and Out. The graphical symbols for EDPNs are shown in Figure 6.1. The symbols for port events could be extended to show the discrete/continuous difference. Alternatively, this can be clarified in a legend that is associated with a diagram, as in Figure 6.2.

Definition: A *thread* is a sequence of transitions in an event-driven Petri Net.

We can always construct the inputs and outputs of a thread from the inputs and outputs of the transitions in the thread. EDPNs are graphically represented in much the same way as ordinary Petri nets, the only difference is the use of triangles

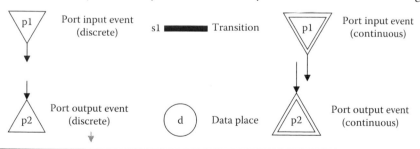

Figure 6.1   Graphical symbols for Event-Driven Petri Nets.

p1: move lever up one position (discrete)
p2: move lever down one position (discrete)
p3: deliver 30 wipes per minute (continuous
    and mutually exclusive)
p4: deliver 60 wipes per minute (continuous
    and mutually exclusive)
d1: lever at INT
d2: lever at LOW
d3: lever at HIGH
s1: transition INT to LOW
s2: transition LOW to HIGH
s3: transition HIGH to LOW
s4: transition LOW to INT

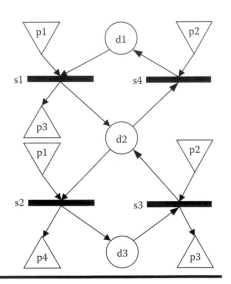

Figure 6.2   EDPN of part of the Windshield Wiper.

for port event places. In the EDPN as shown in Figure 6.2 there are four transitions, s1, s2, s3, and s4, two port input events, p1 and p2, two port output events, p3 and p4, and four data places (states), d1, d2, and d3. This EDPN corresponds to a portion of the finite state machine developed for the Windshield Wiper.

6.1.1 Transition Enabling and Firing

Markings for an EDPN are more complicated because we want to be able to deal with event quiescence.

Definition: A *marking M of an EDPN* (*P, D, T, In, Out*) is a sequence M = <m1, m2,...> of p-tuples, where p = k + n, and k and n are the number of elements in the sets P and D, respectively. Individual entries in a p-tuple indicate the number of tokens in the event or data place.

Definition: A *marked EDPN is a six-tuple* (*P, D, T, In, Out, M*) where P, D, T, In, and Out are as in an unmarked EDPN, and M is a marking sequence.

Token movement and creation in an EDPN requires closer attention. There are only two ways a data place can be marked: either by an initial marking, or by firing a transition. The only way an output event can be marked is by firing a transition. Other than an initial marking, the only way to mark an input event is by choices made by the person who is executing the EDPN. This captures the event-driven nature of event-driven systems exactly. Token duration also requires a more detailed view. Tokens in data places and states are as they were in ordinary Petri nets. Tokens in events are different, partly because events can be either discrete or continuous.

Definition: A *discrete event*, either input or output, is one that has a very short, possibly instantaneous, duration.

Definition: A *continuous event*, either input or output, is one that has a measurable duration.

In the Garage Door Controller, the input events are all discrete, and the output events (motor) are all continuous and mutually exclusive.

How long does a continuous output event remain marked? This is very dependent on the application, but there are two clear cases:

1. When a continuous output event occurs as a result of a transition firing that leaves the net in a state of event quiescence.
2. When continuous output events are mutually exclusive.

In the event quiescence case (case 1), we can think of the duration of the continuous output event as the duration of the event quiescent interval. When output events are mutually exclusive, the duration reduces to the next time a transition causes a different member of the mutually exclusive set of events.

When we make EDPN markings, we will put the input event places first, followed by the data places and then the output event places. An EDPN may have any number of marking sequences; each corresponds to an execution of the net. Table 6.1 shows a sample marking sequence of the EDPN in Figure 6.2.

The rules for transition enabling and firing in an EDPN are exact analogs of those for traditional Petri nets; a transition is enabled if there is at least one token in each input place, and when an enabled transition fires, one token is removed from each of its input places, and one token is placed in each of its output places. The "Time Step" column in Table 6.1 needs some explanation. Recall our refined definition of states in a finite state machine: an interval of time in which a certain proposition is true. The same definition applies to data places in an EDPN when they refer to states or conditions.

Definition: A transition in an Event Driven Petri Net is *enabled* if there is at least one token in each of its inputs (places or events).

Definition: When an enabled Event Driven Petri Net transition *fires*, one token is removed from each of its inputs, and one token is added to each of its outputs.

One important difference between EDPNs and traditional Petri nets is that event quiescence can be broken by creating a token in a port input event place. In traditional Petri nets, when no transition is enabled, we say that the net is deadlocked. In EDPNs, when no transition is enabled, the net may be at a point of event quiescence. (Of course, if no event occurs, this is the same as deadlock.) Event quiescence occurs five times in the thread in Table 6.1: at steps m0, m2, m4, m6, and m8.

Table 6.1   A Marking of the EDPN in Figure 6.2

Time Step	<p1, p2, d1, d2, d3, p3, p4>	Description
m0	<0, 0, 1, 0, 0, 0, 0>	Initial condition, in state d1
m1	<1, 0, 1, 0, 0, 0, 0>	Event p1 occurs, s1 is enabled
m2	<0, 0, 0, 1, 0, 1, 0>	Result of s1 firing, p3 occurs, in state d2
m3	<1, 0, 0, 1, 0, 1, 0>	Event p1 occurs, s2 is enabled
m4	<0, 0, 0, 0, 1, 0, 1>	Result of s2 firing, p4 occurs, in state d3, p3 is unmarked
m5	<0, 1, 0, 0, 1, 0, 1>	Event p2 occurs, s3 is enabled
m6	<0, 0, 0, 1, 0, 1, 0>	Result of s3 firing, p3 occurs, in state d2, p4 is unmarked
m7	<0, 1, 0, 1, 0, 1, 0>	Event p2 occurs, s4 is enabled
m8	<0, 0, 1, 0, 0, 0, 0>	Result of s4 firing, no output event, in state d1*

* Small problem here—p3 remains marked.

The individual members in a marking sequence can be thought of as snapshots of the executing EDPN at discrete points in time; these members are alternatively referred to as time steps, p-tuples, or marking vectors. This lets us to think of time as an ordering that allows us to recognize "before" and "after." If we attach instantaneous time as an attribute of port events, data places, and transitions, we obtain a much clearer picture of thread behavior. One awkward part to this is how to treat tokens in port output events. Port output places always have outdegree = 0 in an ordinary Petri net, so there is no way for tokens to be removed from a place with a zero outdegree. If the tokens in a port output event place persist, this suggests that the event occurs indefinitely. The output events p3 and p4 are mutually exclusive, so the convention we used for ordinary Petri nets can apply here. Once p4 is marked, p3 must be unmarked. (Another possibility is to remove tokens from a marked output event place after one time step; this works reasonably well, as shown in Table 6.1.)

6.1.2 Conventions

Because ordinary Petri nets are a special case of EDPNs, the conventions in Section 5.1.2 also apply to EDPNs. The conventions are repeated here, but the explanations in Chapter 5 are omitted.

Convention 1: Regardless of how many transitions are enabled, only one transition at a time can be fired.

Convention 2: Places may be inputs to more than one transition; symmetrically, places may be outputs of more than one transition.

Convention 3: Events are either inputs to transitions, or outputs of transitions; not both.

Convention 4: Transitions must have at least one input (event or place) and at least one output (again, event or place).

6.1.3 Nongraphical Representations

As we saw for ordinary Petri nets, EDPNs can have either textual or database representations. Both forms are useful, and both relieve the scale-up problem of EDPN drawings, at the expense of loss of visual flow information.

6.1.3.1 Textual Representation

The textual template we had in Chapter 5 for ordinary Petri nets requires only a slight modification to be used for EDPNs.

Transition <transition Id> <transition name>
    has input places (repeat for each input place):
        <place 1> <place name> with marking <number of tokens>
    and has output places (repeat for each output place):
        <place 1> <place name> with marking <number of tokens>
    has input events (repeat for each input event):
        <event 1> <event name> with marking <number of tokens>
    and has output events (repeat for each output event):
        <event 1> <event name) with marking <number of tokens>

Here is the completed EDPN template for transition s1 in Figure 6.2.

Transition <s1> <transition INT to LOW>
    has input places (repeat for each input place):
        <d1> <lever at INT > with marking <1 token>
    and has output places (repeat for each output place):
        <d2> <lever at LOW> with marking <0 tokens>
    has input events (repeat for each input event):
        <p1> <move lever up one position> with marking <0 tokens>
    and has output events (repeat for each output event):
        <p3> <deliver 30 wipes per minute> with marking <0 tokens>

6.1.3.2 Database Representation

The E/R model in Figure 6.3 contains all the relations needed to fully describe an EDPN. All relations are many-to-many with optional participation, that is, their UML min/max descriptions are all (0...n).

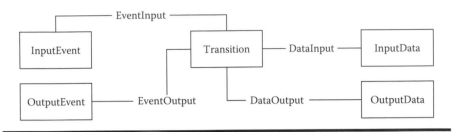

Figure 6.3   E/R model of an EDPN database.

As was the case for ordinary Petri nets, the information in this populated EDPN database is both necessary and sufficient to recreate (except for spatial placement) the EDPN in Figure 6.2. The populated entities and relations in the database for Figure 6.2 are given in Tables 6.2 through 6.4.

Table 6.2   Populated EDPN Place and Transition Entities for Figure 6.2

Place (State)	Name	Tokens
d1	Lever at INT	0
d2	Lever at LOW	0
d3	Lever at HIGH	0

Transition	Name
s1	Transition INT to LOW
s2	Transition LOW to HIGH
s3	Transition HIGH to LOW
s4	Transition LOW to INT

Table 6.3   Populated EDPN Event Entities for Figure 6.2

Input Event			Output Event		
Event	Name	Tokens	Event	Name	Tokens
p1	Move lever up one position	0	p3	Deliver 30 w.p.s.	0
p2	Move lever down one position	0	p3	Deliver 60 w.p.s.	0

Table 6.4 Populated EDPN Relations for Figure 6.2

EventInput		EventOutput		DataInput		DataOutput	
Event	Transition	Event	Transition	Place	Transition	Place	Transition
p1	s1	p3	s1	d1	s1	d1	s4
p1	s2	p4	s2	d2	s2	d2	s1
p2	s3	p3	s3	d2	s4	d2	s3
p2	s4			d3	s3	d3	s2

6.2 Technique

Here are some general hints for using Event Driven Petri Nets:

1. Use transitions to represent actions
2. Use places to represent any of the following:
   - Data
   - Pre- and postconditions
   - States
   - Messages
3. Use events to represent
   - Port input events (including passage of time)
   - Port output events
4. Use the Input relationship to represent prerequisites and inputs to actions.
5. Use the Output relationship to represent consequences and outputs of actions.
6. Use markings to represent "states" of a net, memory, or counters.
7. Tasks can be considered to be individual transitions or subnets.
8. Subsets of input data places to a transition can be used to define a context.

Since ordinary Petri nets are a special case of Event Driven Petri Nets, many of the behavioral issues described in Chapter 5 are identical in the Event Driven Petri Net form. The following issues therefore need no special description (see Chapter 5 for complete definitions):

- Sequence, selection, and repetition
- Enable, disable, and activate
- Trigger
- Suspend, resume, and pause
- Conflict and priority
- Mutual exclusion
- Synchronization

The remaining behavioral primitives (patterns), Context-Sensitive Input events, Multiple Cause Output Events, and Event Quiescence, are discussed here.

6.2.1 Context-Sensitive Input Events

The event (e) in Figure 6.4 is a context-sensitive input event to transitions s1 and s2. Notice that s1 and s2 are in Petri net conflict with respect to event (e). Whichever context (place) is marked determines the "meaning" of (e). In good practice, the contexts of context-sensitive input events should be mutually exclusive. Similarly, the event (o) is an output event that occurs for multiple causes.

6.2.2 Multiple-Cause Output Events

The output event in Figure 6.4 can be caused either by s3 or by s4. With no history, once event o occurs, its cause is not known. This characterizes many field trouble reports where an unexpected output occurs, and its normal cause is not the reason.

6.2.3 Event Quiescence

Transitions s1 and s2 in Figure 6.5 are disabled until event (e) occurs. If (e) never occurs, this would be deadlock. This is the event quiescence of event-driven systems, where nothing happens until an event occurs.

6.2.4 Event-Driven Petri Net Engines

Because Event-Driven Petri Nets are an extension of ordinary Petri nets, there are only a few differences between their respective engines. Given an Event-Driven Petri

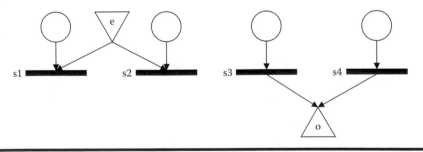

Figure 6.4   Context-sensitive input events and multiple-context output events.

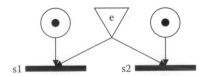

Figure 6.5   Event quiescence.

Net, its possible execution tables indicate how its engine will operate. (See Table 6.6 for a short execution table for the Windshield Wiper problem.)

An EDPN engine needs a definition of all input events, output events, places, and the input and output functions relating these to transitions, and an initial marking, similar to the use case in Table 6.6. Then the engine (and the user) proceeds as follows:

1. The engine determines a list of all enabled transitions.
2.1. If no transition is enabled, the net is either at a point of event quiescence or it is deadlocked.
2.2 If an input event can break event quiescence, the user causes (symbolically) an input event to occur. If more than one input event can end event quiescence, the user chooses which event to cause.
2.3. If the EDPN is deadlocked, the engine ends its execution.
3.1. If only one transition is enabled, it is automatically fired, and a new marking is generated.
3.2. If more than one transition is enabled, the user must select one to fire, and then a new marking is generated. (Note that the user resolves any Petri net conflict.)
3.3. If transition firing causes an output event, that event is marked until the next execution step.
4. The engine returns to step 1 and repeats until either the user stops or the net is deadlocked.

There are seven levels of Event-Driven Petri Net execution; they correspond exactly to execution levels for ordinary Petri nets and for Statecharts.

Level 1 (Interactive): The user provides an initial marking, and then directs transition firing, as described above.
Level 2 (Burst mode): If there is a chain of steps in which only one transition is enabled, the whole chain is fired.
Level 3 (Predetermined): An input script marks places and directs execution.
Level 4 (Batch mode): A set of predetermined scripts is executed.
Level 5 (Probabilistic): Similar to the interactive mode, only conflicts are resolved using transition firing probabilities (*e.g.,* at step 2.3).
Level 6 (Traffic mix): A demographic set of batch scripts, executed in random order.
Level 7 (Exhaustive): For a system with no loops, execute all possible "threads." Exhaustive execution is more sensibly done with respect to the reachability tree of a Petri net. If a system has loops, the use of condensation graphs can produce a loop-free version. This is computationally intense, as a given initial marking may generate many possible threads, and the process should be repeated for many initial markings.

6.2.5 Deriving Test Cases from an Event-Driven Petri Net

Just as Petri nets inherit the advantages of finite state machines, Event-Driven Petri Nets inherit all the advantages of ordinary Petri nets. The big difference, indeed the motivating factor, is that EDPNs deal explicitly with events, and offer the following additional advantages:

- Events are explicitly modeled.
- Context-sensitive input events are easily identified.
- As with flowcharts, finite state machines, and ordinary Petri nets, paths are candidate test cases.

Limitations persist in Event-Driven Petri Nets:

- The convention for unmarking a continuous output event is awkward, but manageable.
- There is no graphical difference between discrete and continuous events.
- EDPNs are clearly overkill for computational applications such as the Insurance Premium problem.

6.3 Examples

6.3.1 Railroad Crossing Gate Controller

Here we revisit the finite state machine model of the Railroad Crossing Gate Controller from Chapter 4 as an EDPN. The EDPN for the full Railroad Crossing Gate Controller is shown in Figure 6.6. All four port events are discrete events.

Port Input Events	Port Output Events	Places
p1. train arrival	p3. lower crossing gate	d1. no trains in the crossing
p2. train departure	p4. raise crossing gate	d2. 1 train in the crossing
		d3. 2 trains in the crossing
		d4. 3 trains in the crossing

The use case in Table 6.5 also serves as both an abstract test case and a "concrete" test case because there are no values needed to change from the abstract version to a concrete version. This is common for systems that are solely event-driven.

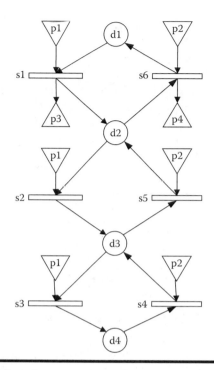

Figure 6.6   Railroad Crossing Gate Controller EDPN.

Table 6.5   Use Case for a Long Path in Figure 6.6

Use Case Name:	Long Path in the Railroad Crossing Gate Controller	
Use Case ID:	WWUC-1	
Description:	Operator moves through a typical sequence of lever and dial events. (w.p.m. is wiper strokes per minute)	
Pre-conditions:	1. d1. No trains in the crossing; (gate is raised)	
Event Sequence:	**Input Event**	**System Response**
	1. p1. Train arrival	2. p3. Lower crossing gate
	3. p1. Train arrival	4. (No output event occurs)
	5. p1. Train arrival	6. (No output event occurs)
	7. p2. Train departure	8. (No output event occurs)
	9. p2. Train departure	10. (No output event occurs)
	11. p2. Train departure	12. p4. Raise crossing gate
Postconditions:	1. d1. No trains in the crossing; (gate is raised)	

6.3.2 *Windshield Wiper Controller*

The EDPN for the full Windshield Wiper Controller is shown in Figure 6.7. The input edges between place d2 (lever at Int) and transitions s11, s12, and s13 (dial positions) are all double-headed connections. Otherwise, these transitions could only fire once. The events and places listed below are used in Figure 6.7 and in Table 6.6.

Input Events (discrete)	Output Events (continuous)	Places
p1: move lever up one position	p10: deliver 0 strokes per minute	d1: lever at OFF
p2: move lever down one position	p11: deliver 6 strokes per minute	d2: lever at INT
p3: move dial up one position	p12: deliver 12 strokes per minute	d3: lever at LOW
p4: move down one position	p13: deliver 20 strokes per minute	d4: lever at HIGH
	p14: deliver 30 strokes per minute	d5: dial at 1
	p15: deliver 60 strokes per minute	d6: dial at 2
		d7: dial at 3

Executing an EDPN is more complex than executing ordinary Petri nets because there are two distinct types of events: discrete and continuous. These relate to marking, the duration of a time step, and event quiescence. The output events (p10–p15) are best understood as continuous, in the sense that they deliver various wiper speeds for the duration of the time step in which they occur. Table 6.6 shows an execution table that could be produced by an EDPN engine. It follows the use case introduced in Chapter 5.

In time step m1, since p1 occurs, we cannot say that the system is event quiescent. But, since p1 is a discrete event (with instantaneous duration), transition s1 is enabled. If we decide that, when only one transition in an EDPN is enabled, it fires immediately, then all of time step m1 is instantaneous. Otherwise, we can show the transition firing as a separate step, and in step m2. In time step m3 (similarly for time steps m6 and m17), since p11 occurs, no transition is enabled. So now we must clarify what we mean by event quiescence.

Choice 1: No transition is enabled, waiting for an input event.
Choice 2: No transition is enabled, a continuous output event is occurring.
Choice 3: The EDPN is truly deadlocked, *that is*, there is no input event that can enable a transition.

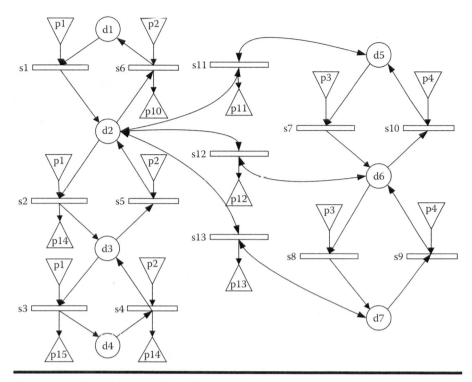

Figure 6.7   Windshield Wiper Controller EDPN.

Table 6.6   EDPN Execution Table for the Windshield Wiper Controller

Time Step	Fired Transition	Input Event	Marked Input Events and Places	Enabled Transitions	Output Event	Marked Output Events and Places	Event Quiescent?
m0			d1, d5	(none)			yes
m1		p1	p1, d1, d5	s1			no
m2	s1		d2, d5	s11		d2, d5	no
m3	s11		d2, d5	s11*	p11	d2, d5, p11	yes
m4		p3	p3, d2, d5	s7, s11		d2,d5	no
m5	s7		d2, d6	s12			no
m6	s12		d2, d6	s12	p12	d2, d6, p12	yes
m7		p1	p1, d2, d6	s2, s12			no

(Continued)

Table 6.6 (*Continued*) EDPN Execution Table for the Windshield Wiper Controller

Time Step	Fired Transition	Input Event	Marked Input Events and Places	Enabled Transitions	Output Event	Marked Output Events and Places	Event Quiescent?
m8	s2		d3, d6		p14	d3, d6, p14	yes
m9		p1	p1, d3, d6	s3		d3, d6	no
m10	s3		d4, d6		p15	d4, d6, p14	yes
m11		p2	p2, d4, d6	s4		d4, d6	no
m12	s4		d3, d6		p14	d3, d6, p14	yes
m13		p3	p3, d3, d6	s8		d3, d6	no
m14	s8		d3, d7			d3, d7	yes
m16		p2	p2, d3, d7	s5		d3, d7	no
m16	s5		d2, d7	s13		d2, d7	no
m17	s13		d2, d7	s13	p13	d2, d7	yes
m18		p2	p2, d2, d7	s6, s13		d2, d7	no
m19	s6		d1, d7		p10	d1, d7	yes

In step m4, transitions s7 and s11 are in conflict. The EDPN is quiescent until the conflict is resolved, but how might this happen? If a customer executes the model, using an EDPN engine, the customer choice is, in a curious sense, also an input.

6.4 Deriving Test Cases from an Event-Driven Petri Net

There is a sense of "reverse inheritance" that applies to the question of deriving test cases from an EDPN: finite state machines are a special case of ordinary Petri nets, which, in turn, are a special case of EDPNs. So when a problem is modeled as an EDPN, all the test case derivation information from finite state machines and ordinary Petri nets is available—and more. Now the EDPN formulation deals directly with events. In Chapter 4 we defined a *system-level finite state machine* (SL/FSM) as one in which paths corresponded directly to system level test cases. This is clearly true for paths in an EDPN. Table 6.7 shows the alignment between use cases, system test cases, and elements in an EDPN. Table 6.7 also shows that the EDPN formulation of an application provides a top-down view, whereas use cases and test cases are both bottom-up views.

Table 6.7 Use Cases, System Test Cases, and EDPN Elements

Use Case	System Test Case	Event-Driven Petri Net
ID	ID	System Name
Name	Name	List of all system elements: input events, output events, date places, and transitions
Narrative description	Test objective	Path description (usually a sequence of transitions)
Preconditions	Preconditions	Data places with indegree = 0
Event Sequence (inputs interleaved with outputs)	Event Sequence (inputs interleaved with outputs)	Inputs to and outputs of transitions
Postconditions	Postconditions	Data places with outdegree = 0
	Pass/fail result (and other test management information)	

When paths in an EDPN are converted to test cases, there is support for the following set of system test coverage metrics:

- Every transition (atomic, system level function)
- Every input event
- Every context-sensitive input event
- Every input event in every context
- Every output event
- Every multiple-cause output event
- Every output event for every cause
- Every precondition
- Every postcondition
- Every path (excluding loops)

6.4.1 The Insurance Premium Problem

There are no events in the Insurance Premium problem so its Event-Driven Petri Net is the same as its ordinary Petri net (see Section 5.4.1). The same test cases from Chapter 5 could be derived from the EDPN formulation.

6.4.2 The Garage Door Controller

The input and output events and states for the EDPN formulation of the Garage Door Controller are nearly identical to those in the ordinary Petri net version, except the places have been renamed. The EDPN formulation resolves the problems indentified in Chapter 5 for the ordinary Petri net formulation. Specifically, we assume that input events can be "created" at any time by the User. This obviates the need for an initial marking that includes every input event, as for ordinary Petri nets. The persistent marking of output events is ameliorated by two conventions:

1. For mutually exclusive output events, a marked output event is unmarked when one of its mutually exclusive events is marked by transition firing, or
2. A marked output event remains marked until the *next* transition is fired.

Input Events	Output Events	States
p1: Control signal	p5: Start drive motor down	d1: Door up
p2: End of down track hit	p6: Start drive motor up	d2: Door down
p3: End of up track hit	p7: Stop drive motor	d3: Door stopped going down
p4: Laser beam crossed	p8: Reverse motor down to up	d4: Door stopped going up
		d5: Door closing
		d6: Door opening

Transitions

    t1: Door up to door closing
    t2: Door closing to door stopped going down
    t3: Door closing to door down
    t4: Door closing to door opening (safety reversal)
    t5: Door stopped going down to door closing
    t6: Door down to door opening
    t7: Door opening to door stopped going up
    t8: Door stopped going up to door opening
    t9: Door opening to door up

Figures 6.8 and 6.9 gradually build up to the full Garage Door Controller EDPN in Figure 6.10. Figure 6.8 also shows how EDPNs can be composed. Development of a full EDPN can be done either in a top-down or in a bottom-up way—the components in Figure 6.9 show part of this process. This is consistent with the "agile model-driven development" proposal by Scott Ambler [Ambler 2004].

The full Garage Door Controller EDPN in Figure 6.10 illustrates a few of the ESML prompts discussed in Chapter 5. The place p5: Door Closing, is an input

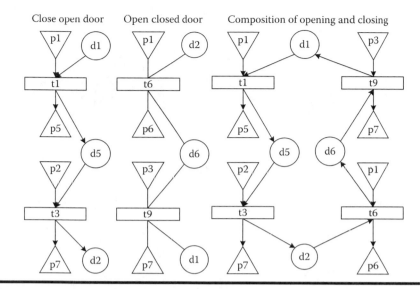

Figure 6.8   Closing and opening garage door, and their composition.

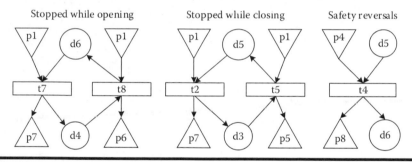

Figure 6.9   Intermediate garage door stops, and safety reversal.

to transitions t2, t3, and t4, so all three are in Petri net *conflict* with respect to place p5. It also *enables* the same three transitions. Finally, it is a point of event quiescence, because once it is marked by either transitions t1 or t5, no transition is enabled. Any of the events p1, p2, or p4 breaks the conflict.

There are a few ESML prompts in Figure 6.11: the firing sequence t2, t5 is a *Pause*: t2 *disables* t4, and t5 *enables* t4. This is exactly how the safety mechanism should work. Notice that firing transition t3 also *disables* transition t4. Event p4 (laser beam crossed) acts like a *Trigger* for t4 when d5 is marked. There are other prompts, but these are the most interesting.

The EDPN in Figure 6.10 still shows the points at which the input events can occur. This is a little artificial; Figure 6.11 shows the same EDPN with single appearances of the input and output events—at the expense of the graphical equivalent of "spaghetti code." Figures 6.12 through 6.20 show the tokens in the marking sequence described in Table 6.8.

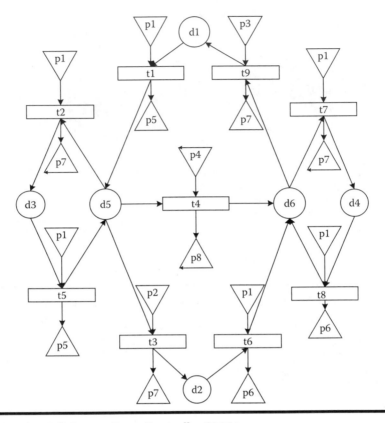

Figure 6.10 Full Garage Door Controller EDPN.

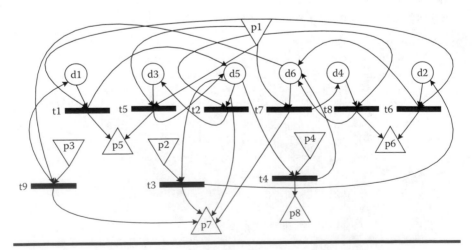

Figure 6.11 Full Garage Door Controller EDPN no repeated events.

Figures 6.12 through 6.20 show the step level information that supports automatic derivation of a system level use case (and test case). The firing sequence displayed in this sequence of figures is part of the longer use/test case in Table 6.10

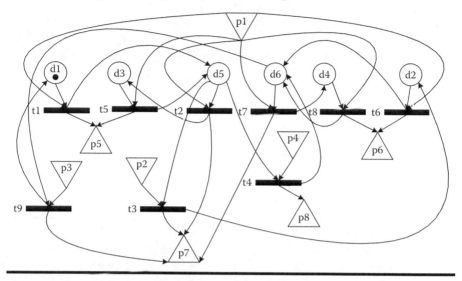

Figure 6.12 Initial marking (door Up). This is the pre-condition for the use/test case. At this point, the system is event quiescent. This diagram also shows the extreme context sensitivity of event p1 the control signal—it can occur in six contexts

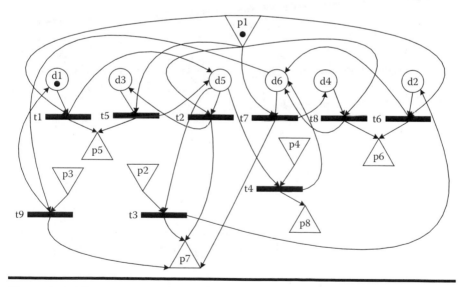

Figure 6.13 Event p1 occurs; transition t1 enabled. Since d1 is marked when p1 occurs, the context sensitivity is resolved, and t1 can fire.

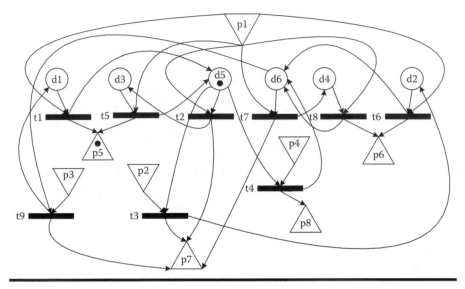

Figure 6.14   After t1 fires, d1 is unmarked, d5 is marked, and p5 occurs. Since p5 is a motor event, it is a continuous output event, and remains marked until some other motor event occurs. At this point, the system is event quiescent.

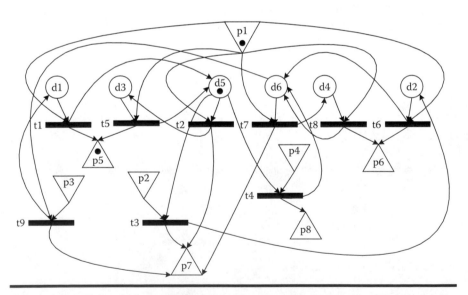

Figure 6.15   Since d5 is marked when p1 occurs, the context sensitivity of event p1 is resolved, thereby enabling t2.

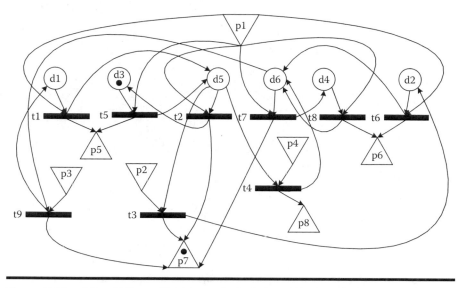

Figure 6.16 After t2 fires, d5 is unmarked, d3 is marked, and p7 occurs. Since p7 is a motor event, it is a continuous output event, and remains marked until some other motor event occurs. This removes the token from continuous output event p5. At this point, the system is event quiescent.

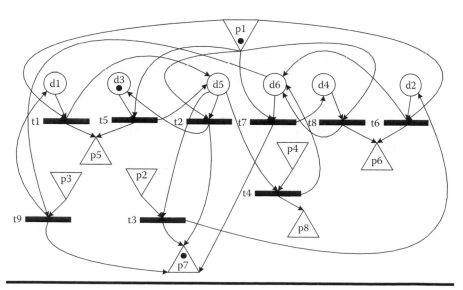

Figure 6.17 Since d3 is marked when p1 occurs, the context sensitivity is resolved, and t5 can fire.

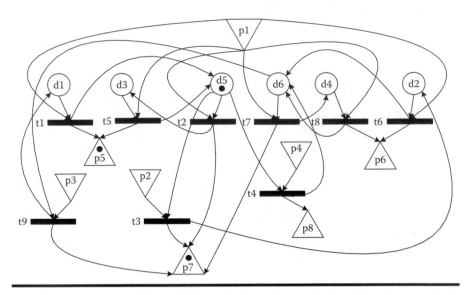

Figure 6.18 Once t5 fires, d5 is marked, p5 is marked, and p7 is unmarked. System is now event quiescent. This marking step is identical to that in Fig. 6.14. Transitions t2, t3, and t4 prompted with an Enable.

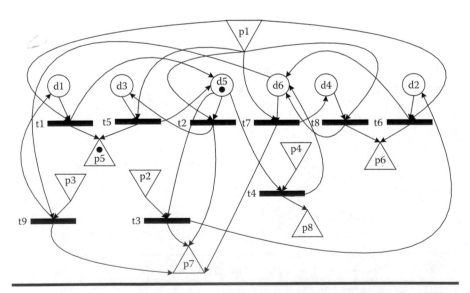

Figure 6.19 Event p2 occurs; since d5 is marked when p2 occurs, the context sensitivity is resolved, and t3 can fire. Output event p5 is still marked.

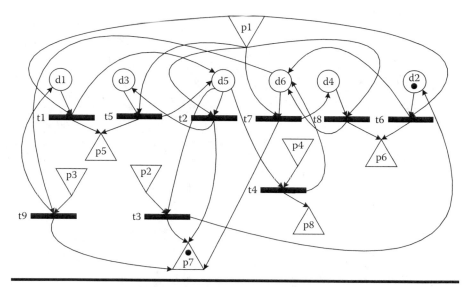

Figure 6.20 Once t3 fires, d2 is marked. p7 is marked, and p5 is unmarked. The garage door is closed at this point of event quiescence.

Table 6.8 Marking Sequence for Door Closing with an Intermediate Stop

Marking Step	Marked Elements	Event Quiescent?	Enabled Transitions	Transition to Be Fired	Figure
0	d1	Yes	None	None	6.12
1	d1, p1	No	t1	t1	6.13
2	d5, p5	Yes	None	None	6.14
3	d5, p1, p5	No	t2, t3, t4	t2	6.15
4	d3, p7	Yes	None	None	6.16
5	d5, p1, p5	No	t5	t5	6.17
6	d5, p5	Yes	None	None	6.18
7	d5, p2, p5	No	t2, t3, t4	t3	6.19
8	d2, p7	Yes			6.20

Once t3 fires, d2 is marked. p7 is marked and p5 is unmarked. The garage door is closed at this point of event quiescence. The firing sequence displayed in this sequence of figures is part of the longer use/test case in Table 6.10.

The paths in Table 6.9 correspond directly to those derived from the decision table formulation in Table 3.15 and to those in Table 5.7 from Chapter 5.

Table 6.10 illustrates the detailed use case for a "soap opera" path (one that is as long as possible). The transition sequence is <t1, t2, t5, t4, t7, t8, t9>. This path covers all but one of the places and all but two of the input events.

Table 6.9 Sample Garage Door Controller Paths

Sample Garage Door Paths (As EDPN Transition Sequences)		
Path	Description	Transition Sequence
1	Normal door close	t1, t3
2	Normal door close with one intermediate stop	t1, t2, t5, t3
3	Normal door open	t6, t9
4	Normal door open with one intermediate stop	t6, t7, t8, t9
5	Closing door with light beam interruption	t1, t4, t9

Table 6.10 Soap Opera Use Case for a Long Path in Figure 6.10

Use Case Name:	Garage Door Controller Soap Opera Use Case (from a Petri net)	
Use Case ID:	GDC-UC-1	
Description:	Use case for the EDPN transition sequence <t1, t2, t5, t4, t7, t8, t9>.	
Preconditions:	1. Garage Door is Up (d1)	
Event Sequence:	Input Event (place)	Output Event (place)
	1. p1: Control signal	2. p5: Start drive motor down
	3. p1: Control signal	4. p7: Stop drive motor
	5. p1: Control signal	6. p5: Start drive motor down
	7. p4: Laser beam crossed	8. p8: Reverse motor down to up
	9. p1: Control signal	10. p7: Stop drive motor
	11. p1: Control signal	12. p6: Start drive motor up
	13. p3: End of up track hit	14. p7: Stop drive motor
Postconditions:	1. Garage door is up (d1)	

6.5 Lessons Learned from Experience

The most important lesson is that, even for examples as small as the Garage Door Controller, Event-Driven Petri Nets do not scale up well. The EDPN database in Figure 6.3 is the best answer to the scale-up problem. Tables 6.11 through 6.13 show the populated entities and relations of the EDPN database for the Garage Door Controller.

The EDPN database is also an excellent answer to the question of Petri net and EDPN composition. Both forms suffer in the drawing mode, but both can be exactly represented in their respective databases, with no loss of information. With the database formulation, the only limit to scaling up is in database storage. The important information easily seen (in small examples) visually can be obtained by well-chosen database queries. For example, input event p1 (the controller signal) is a context-dependent event with six contexts. This can be seen (with a little difficulty) in the figures, and can also be learned from a projection on the EventInput relation that is joined to the DataInput relation.

Figure 6.21 shows the spaghetti-like EDPN diagrams that quickly result, even for small examples.

Table 6.11   Garage Door Controller EDPN Events and Places

Input Event		Output Event		Data Place	
Event	Name	Event	Name	Place	Name
p1	Control signal	p5	Start drive motor down	d1	Door up
p2	End of down track hit	p6	Start drive motor up	d2	Door down
p3	End of up track hit	p7	Stop drive motor	d3	Door stopped going down
p4	Laser beam crossed	p8	Reverse motor down to up	d4	Door stopped going up
				d5	Door closing
				d6	Door opening

Table 6.12 Garage Door Controller EDPN Transitions

Transitions	
Transition	Name
t1	Door up to door closing
t2	Door closing to door stopped going down
t3	Door closing to door down
t4	Door closing to door opening (safety reversal)
t5	Door stopped going down to door closing
t6	Door down to door opening
t7	Door opening to door stopped going up
t8	Door stopped going up to door opening
t9	Door opening to door up

Table 6.13 Garage Door Controller EDPN Populated Relations

EventInput		EventOutput		DataInput		DataOutput	
Event	Transition	Event	Transition	Place	Transition	Place	Transition
p1	t1	p5	t1	d1	t1	d1	t9
p1	t2	p5	t5	d2	t6	d2	t3
p1	t3	p6	t6	d3	t5	d3	t2
p1	t5	p6	t8	d4	t8	d4	t7
p1	t7	p7	t2	d5	t2	d5	t1
p1	t8	p7	t3	d5	t3	d5	t5
p2	t3	p7	t7	d5	t4	d6	t6
p3	t9	p7	t9	d6	t7	d6	t8
p4	t4	p8	t4	d6	t9	d6	t4

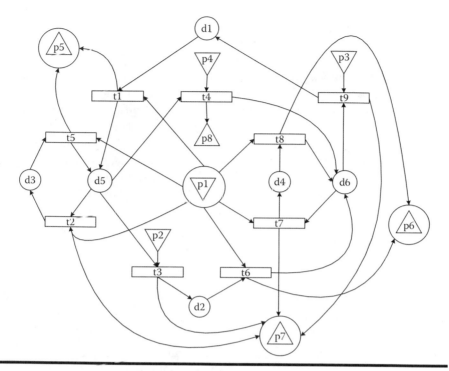

Figure 6.21   Garage Door EDPN drawn with single event placement.

6.6 Advantages and Limitations

Event-Driven Petri nets inherit all of the advantages of ordinary Petri nets, but the addition of event places removes a few limitations. The primary advantage of EDPNs is their explicit modeling of events. This, in turn, permits the recognition of event-related issues such as event quiescence, context-sensitive input events, and multiple-context output events. Similar to ordinary Petri nets, they constitute a bottom-up approach (appropriate for agile development), and can easily be composed into larger nets. As just noted, EDPN engines provide extensive analytic capabilities very early in a project, so they support the executability goal described in Chapter 1. Finally, they allow the recognition and analysis of situations such as starvation (conflict in which one transition always fires), deadlock, and mutual exclusion. Petri nets are probably best used for an extremely "close look" and important portions of a system.

The biggest problem with Petri nets is that the graphical version does not scale up well. This is/can be mitigated by using a database with relations for the information in an Event-Driven Petri net, and then developing sets of useful queries. An E/R model of such a database is given in Figure 6.3. The database possibility completely resolves the scale-up problem but the resolution is at the expense of visual clarity. Closely related to this, most users find that they use a key (or legend) to give short names to Petri net places and transitions, thereby improving communication.

Also, they do not deal with selection in a convenient way: outcomes of a decision must be input places to two transitions that are in conflict. Another disadvantage (resolved by Event Driven Petri Nets) is that they do not represent event-driven systems well. In an execution table, or with an engine, there is no way to mark a place that corresponds to an input event. Also, once places that correspond to output events are marked, they remain marked. Finally, Petri nets are dismal at describing mathematical calculations. Technically, Event-Driven Petri nets do not represent concurrency—only one transition can be fired at a time. As with ordinary Petri nets, this can be circumvented by assigning subnets to separate devices. This deficiency will be elegantly resolved with Statecharts.

This chapter concludes with the continuing comparison of behavioral issues modeled by EDPNs (Table 6.14).

Table 6.14   Representation of Behavioral Issues with EDPNs

Behavioral Issue	Represented in EDPNs?
Sequence	Yes
Selection	Yes, but awkward
Repetition	Yes
Enable	Yes
Disable	Yes
Trigger	Yes
Activate	Yes
Suspend	Yes
Resume	Yes
Pause	Yes
Priority	Yes
Mutual exclusion	Yes
Concurrent execution	No
Deadlock	Yes
Context-sensitive input events	Yes
Multiple-context output events	Yes
Asynchronous events	Yes
Event quiescence	Yes

References

[Ambler 2004]

Ambler, Scott, Agile Model-Driven Development, http://agilemodeling.com/essays/, also in *The Object Primer 3rd Edition: Agile Model Driven Development with UML 2*, Cambridge University Press, 2004.

[DeVries 2013]

DeVries, Byron, *Mapping of UML Diagrams to Extended Petri Nets for Formal Verification*, Master's Thesis, Grand Valley State University, Allendale, MI, 2013.

Chapter 7

Statecharts

David Harel had two goals when he developed Statecharts: he wanted to devise a visual notation that combined the ability of Venn diagrams to express hierarchy and the ability of directed graphs to express connectedness [Harel 1988]. Taken together, these capabilities provide an elegant answer to the "state explosion" problem of ordinary finite state machines. The result is a highly sophisticated and very precise notation that is supported by commercially available CASE tools, notably the StateMate system from IBM. Statecharts are now the control model of choice for the Unified Modeling Language (UML) from the Object Management Group. (See www.omg.org for more details.) Much of Section 7.1 is taken directly from [Jorgensen 2009].

7.1 Definition and Notation

Definition: A *Statechart* is a hierarchical directed graph (S, T, R, In, Out), where

- S is a set of states
- T is a set of transitions
- R is a set of orthogonal regions
- In is a set of inputs
- Out is a set of outputs

  Subject to the following conventions:

- States may contain substates
- States may contain orthogonal regions

- Transitions are visually defined by their initial and final points on the boundaries (contours) of states
- Transitions are caused by inputs in In
- Transitions cause Outputs in Out

Elements of In and Out can be any combination of port events, data values, or conditions involving any of these.

Harel uses the methodology neutral term "blob" to describe the basic building block of a Statechart. Blobs can contain other blobs the way that Venn diagrams show set containment. Blobs can also be connected to other blobs with edges in the same way that nodes in a directed graph are connected. As Harel intended, we can interpret blobs as states, and edges as transitions. The full StateMate system supports an elaborate language that defines how and when transitions occur. Statecharts are executable, in a much more elaborate way than ordinary finite state machines. Executing a Statechart requires a notion similar to that of Petri net markings.

In Figure 7.1, blob A contains two blobs, B and C, and they are connected by edges. Blob A is connected to blob D by two edges. The "initial blob" of a Statechart is indicated by an edge that has no source blob. When blobs are nested within other blobs, the same indication is used to show the lower level initial blob. In Figure 7.1, blob A is the initial blob, and when it is entered, the lower level blob B is also entered. When a blob is entered, we can think of it as active, in a way very analogous to a marked place in a Petri net. If a blob contains other blobs, as blob A does, the edge "refers" to all subblobs. Thus the edge from A to D means that the transition can occur either from blob B or from blob C. The edge from D to blob A, as in Figure 7.1, as B is indicated as the initial blob means that the transition is really from blob D to blob B. This convention greatly reduces the tendency of finite state machines to look like "spaghetti code." From here on, in this chapter, we will just refer to states rather than the more neutral term "blob."

There is one specialized difference between Statechart states and the circles of a Venn diagram. In a Venn diagram, any point within a circle is assumed to be an element of the set represented by the circle. The same is not true for state inclusion. An included state (a substate) must have a unique contour. In Figure 7.2, state B is a substate of state A, but state C is problematic because its contour touches the contour of state A. Figure 7.2 demonstrates the reason for the "unique contour"

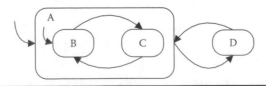

Figure 7.1   Blobs, contours, and edges in a Statechart.

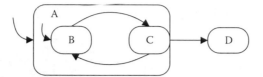

Figure 7.2 Convention of state contours.

convention: as the contour of blob C coincides with the contour of blob A, there is no way to know which state, A or C, is the source state of the transition to state D.

The Statechart language is quite complex, and represents a major extension to that of finite state machines. One basic idea is that Statecharts now have memory— recall that finite state machines have no memory. This one change greatly enhances the expressive power of a Statechart model. Transitions are still caused by events and conditions (as described in Tables 7.1 through 7.3).

Table 7.1 Statechart Language Elements for Events

Event	Occurs When:
en(S)	State S is entered.
ex(S)	State S is exited.
entering(S)	State S is entered.
exiting(S)	State S is exited.
st(A)	Activity A is started.
sp(A)	Activity A is stopped.
ch(V)	The value of data item expression V is changed.
tr(C)	The value of condition C is set to TRUE (from FALSE).
fs(C)	The value of condition C is set to FALSE (from TRUE).
rd(V)	Data item V is read.
wr(V)	Data item V is written.
tm(E, N)	N clock units passed from last time event E occurred.
E(C)	E has occurred and condition C is TRUE.
not E	E did not occur.
E1 and E2	E1 and E2 occurred simultaneously.
E1 or E2	E1 or E2, or both occurred.

Table 7.2   Statechart Language Elements for Conditions

Condition	TRUE When:
in(S)	System is in state S.
ac(A)	Activity A is active.
hg(A)	Activity A is suspended.
EXP1 R EXP2	The value of the expression EXP1 and EXP2 satisfy the relation R. When expressions are numeric, R may be: =, /=, > or <. When they are strings, R may be: = or/=.
not C	C is not TRUE.
C1 and C2	Both C1 and C2 are TRUE.
C1 or C2	C1 or C2, or both are TRUE.

Table 7.3   Statechart Language Elements for Actions

Action	Performs:
E	Generate the event E.
tr!(C)	Assign TRUE to the condition C.
fs!(C)	Assign FALSE to the condition C.
V:=EXP	Assign the value of EXP to the data item V.
st!(A)	Activate the activity A.
sp!(A)	Terminate the activity A.
sd!(A)	Suspend the activity A.
rs!(A)	Resume the activity A.
rd!(V)	Read the value of data item V.
wr!(V)	Write the value of data item V.

The last aspect of Statecharts we will discuss is the notion of concurrent or orthogonal regions. In Figure 7.3, the dotted line in state D is used to show that state D really refers to two concurrent states, E and F. (Harel's convention is to move the state label of D to a rectangular tag on the contour of the state.) We can (should!) think of E and F as parallel machines (or better, devices) that execute

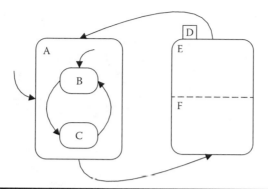

Figure 7.3 Concurrent states.

concurrently. Because the edge from state A terminates on the contour of state D, when that transition occurs, both machines E and F are active (or marked, in the Petri net sense). Each orthogonal region must have its own default entry.

7.2 Technique

Here are some general hints for using Statecharts:

1. Use transitions to represent state/substate changes.
2. Annotate transitions with the language capabilities in Tables 7.1 through 7.3.
3. Use states to represent any of the following:
   - Data
   - Pre- and postconditions
4. Use inputs and outputs to represent
   - Port input events (including passage of time)
   - Port output events
   - Data conditions
5. Use orthogonal regions to represent separate devices or concurrent processing.

7.2.1 Communication with the Broadcasting Mechanism

Figure 7.4 is taken directly from David Harel's seminal paper in 1988 [Harel 1988]. He uses this example to describe Statechart execution. It has three concurrent regions, labeled here as A, D, and H, and these are always active. On entry to the overall (not named) chart, the initially active substates are B, F, and J. We will follow an abstract scenario to see the effect of events on active states in Table 7.4.

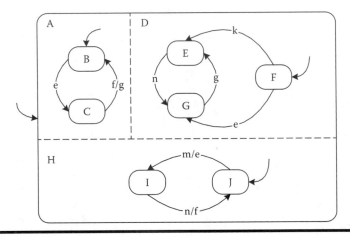

Figure 7.4   Broadcasting Statechart from Harel (1988).

Table 7.4   Execution Table for Harel's Broadcasting Example

Step	Active Substate(s)	Input Event	Output Events	Next Substate(s)
0	B, F, J	k	(None)	E
1	B, E, J	m	E	I
2	B, E, I	e	(None)	C
3	C, E, I	n	f	G, J
4	C, G, J	f	g	B, G, J
5	B, G, J	g	(None)	B, E, J

7.2.2 *Statechart Engines*

Input to the original StateMate system is a Statechart. There are seven levels of Statechart execution. A Statechart engine would produce an execution table similar to the one in Table 7.5, which traces the same use case scenario that was used in Chapters 4, 5, and 6. Statechart execution has a notion of time steps, which are time intervals between input events. In Table 7.5, the first time step is split into three stages to show the effect of broadcasting. These stages are compressed into one in the remaining time steps. The Statechart notation does not provide for direct expression of output events—but they are easily shown as states in a separate concurrent region. In the original StateMate system, outputs on transitions referred to actions which, in turn, generated output events.

Table 7.5 Execution Table for the Saturn Windshield Wiper Controller

Step	Active States	Input Events	Next States	Output (Events)
1.1	Lever. Off, Dial.1, Wiper.0wpm	e1: Move lever up one position	Lever. Int, Dial.1	In(Int), In(1)
1.2	Lever. Int, Dial.1	In(Int)	Lever. Int, Dial.1, Wiper. Int	
1.3		In(1)	Lever. Int, Dial.1, Wiper. Int.6wmp	
2	Lever. Int, Dial.1, Wiper. Int.6wmp	e3: Move dial up one position	Lever. Int, Dial.2, Wiper. Int.12wmp	In(Int), In(2)
3	Lever. Int, Dial.2, Wiper. Int.12wmp	e1: Move lever up one position	Lever. Low, Dial.2, Wiper.30wmp	In(Low), In(2)
4	Lever. Low, Dial.2, Wiper30wmp	e1: Move lever up one position	Lever. High, Dial.2, Wiper.60wmp	In(High), In(2)
5	Lever. High, Dial.2, Wiper.60wmp	e2: Move lever down one position	Lever. Low, Dial.2, Wiper.30wmp	In(Low), In(2)
6	Lever. Low, Dial.2, Wiper.30wmp	e3: Move dial up one position	Lever. Low, Dial.3, Wiper.30wmp	In(Low), In(3)
7	Lever. Low, Dial.3, Wiper.30wmp	e2: Move lever down one position	Lever. Int, Dial.3, Wiper. Int.20wmp	In(Int), In(3)
8	Lever. Int, Dial.3, Wiper. Int.20wmp	e2: Move lever down one position	Lever. Off, Dial.3, Wiper.0wmp	In(Off), In(3)

Level 1 (Interactive): The user provides initial conditions, and then directs execution by providing input events.

Level 2 (Burst mode): If there is a chain of steps in which only one transition is enabled, the whole chain is fired. If multiple transitions are enabled in separate concurrent regions, they all execute.

Level 3 (Predetermined): An input script marks places and directs execution.

Level 4 (Batch mode): A set of predetermined scripts is executed.

Level 5 (Probabilistic): Similar to the interactive mode, only conflicts are resolved using transition firing probabilities (e.g., at step 2.3).

Level 6 (Traffic mix): A demographic set of batch scripts, executed in random order.

Level 7 (Exhaustive): For a system with no loops, execute all possible "threads." Exhaustive execution is more sensibly done with respect to the reachability tree of a Petri net. If a system has loops, the use of condensation graphs can produce a loop-free version. This is computationally intense, as a given initial marking may generate many possible threads, and the process should be repeated for many initial markings.

7.2.3 Deriving Test Cases from a Statechart

Since Types I, II, and III of UML Statecharts have been shown to be equivalent to Swim Lane Event-Driven Petri Nets, many of the advantages of EDPNs also apply to Statecharts. The StateMate commercial product has supported automatic generation of system test cases since the early 1990s. The advantages, particularly commercial tool support, are obvious. Some limitations remain however:

- The concurrent regions of Statecharts show potential concurrency, but only one event can occur at a time, and only one transition can occur at a time. The instantaneous convention of the broadcasting mechanism provides almost simultaneous transition firing.
- Events are not as obvious as they are in EDPNs.
- There is no easy way to recognize context-sensitive input events.

7.3 Examples

7.3.1 Railroad Crossing Gate Controller

We continue the models of the Railroad Crossing Gate controller started in Chapter 4. As Statecharts do not deal directly with events, the notation from Chapter 6 (Event-Driven Petri Nets) reverts to the event/action style from finite state machines.

Input Events	Output Actions	States
e1. Train arrival	a1. Lower crossing gate	s1. No trains in the crossing
e2. Train departure	a2. Raise crossing gate	s2. 1. Trains in the crossing
		s3. 2. Trains in the crossing
		s4. 3. Trains in the crossing

The Statechart in Figure 7.5 contains an example of hierarchy, but it is really unnecessary. We could have made it just like the finite state machine in Chapter 4. If some event pertained to each of the states s2, s3, and s4, the "Intersection Occupied" state could be useful. As we might expect, the use case for the long path (as in Chapters 4, 5, and 6) is nearly identical to the earlier versions (Table 7.6).

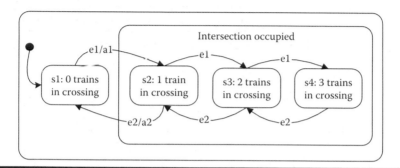

Figure 7.5   Statechart for the Railroad Crossing Gate Controller.

Table 7.6   Use Case for a Long Path in Figure 7.5

Use Case Name:	Long Path in the Railroad Crossing Gate Controller	
Use Case ID:	RRX-UC-1	
Description:	A train arrives at an empty intersection. After the gate is lowered, two more trains arrive, and later, depart. When the last train departs, the gate is raised.	
Preconditions:	1. s1. 0 trains in the crossing; (Gate is raised)	
Event Sequence:	**Input Event**	**System Response**
	1. e1. Train arrival	2. a1. Lower crossing gate
	3. e1. Train arrival	4. (No output event occurs)
	5. e1. Train arrival	6. (No output event occurs)
	7. e2. Train departure	8. (No output event occurs)
	9. e2. Train departure	10. (No output event occurs)
	11. e2. Train departure	12. a2. Raise crossing gate
Postconditions:	1. s1. 0 Trains in the crossing; (Gate is raised)	

7.3.2 *Windshield Wiper Controller*

The input events and output actions are as they were for the finite state machine example in Chapter 4 (Figure 7.6).

Input Events	Output Actions
e1: Move lever up one position	a1: Deliver 0 wipes per minute
e2: Move lever down one position	a2: Deliver 6 wipes per minute
e3: Move dial up one position	a3: Deliver 10 wipes per minute
e4: Move dial down one position	a4: Deliver 20 wipes per minute
	a5: Deliver 30 wipes per minute
	a6: Deliver 60 wipes per minute

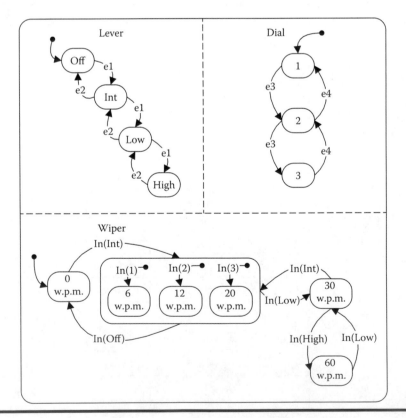

Figure 7.6 Saturn Windshield Wiper Controller Statechart.

Notice the use of the "in state" language element (in(S)) to resolve the destination of transitions into the intermittent state. This is the Statechart mechanism to support communicating finite state machines (hinted at in Chapter 4).

7.4 The Continuing Problems

7.4.1 The Insurance Premium Problem

As in the other transition-based models, there is no reason to model the Insurance Premium Problem as a Statechart. (To do so would be yet another example of model overkill.)

7.4.2 The Garage Door Controller

Figure 7.7 is the finite state machine for the Garage Door Controller from Chapter 4. In Figure 7.8, the original actions are renamed as events that will be broadcast to other orthogonal regions of the Garage Door Controller Statechart. We will borrow the notion of tokens from Petri nets to indicate which states are "active." It is only necessary to use a token to mark the most deeply nested state because each of its super states must also be active.

Input Events	Output Events
e1: Control signal	a1: Start drive motor down
e2: End of down track hit	a2: Start drive motor up
e3: End of up track hit	a3: Stop drive motor
e4: Laser beam crossed	a4: Reverse motor down to up

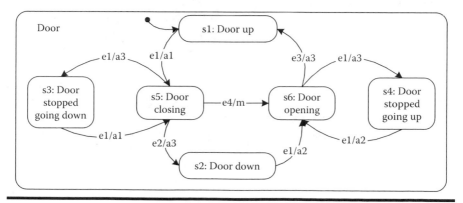

Figure 7.7   Finite state machine for the Garage Door Controller.

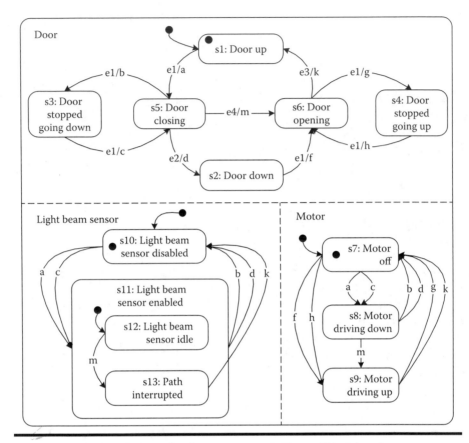

Figure 7.8 Initial *Marking* for the Garage Door Controller Statechart.

In the initial marking in Figure 7.8, states s1, s7, and s10 are all active. Notice that the "actions" from the finite state machine model in Figure 7.7 are now single letters to show the broadcasting communication among concurrent orthogonal regions.

Figure 7.9 introduces another graphical convention—the darkened edges indicate the event(s) that occur, and the consequent broadcasting to the other orthogonal regions. In this figure, event e1 causes a transition from the initial state, s1 to state s5: Door Closing. The broadcast action a causes a transition from s7: motor off to the state s8: motor driving down. At the same time, action a acts as an enable prompt in the Light Beam Sensor region. The transition

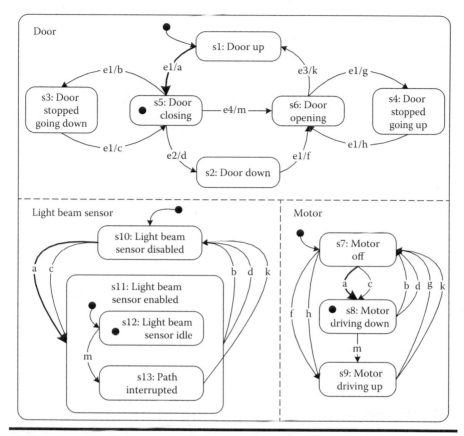

Figure 7.9 Event e1 causes broadcast output a to motor and light beam sensor.

moves to state s12: light beam sensor idle, which is a substate of s11: light beam sensor enabled. As a result of event e1, the active states are s5, s8, and s12, and the system is event quiescent. At this point, any of three events could occur in the door orthogonal region, e1, the control signal, e2, the end of down track reached, or e4, the light beam interruption. There is a problem with time in state s5. The state describes the door closing, and eventually, it will reach to e3: end of down track sensor, so it is probably more accurate to say that the system is temporarily event quiescent. (My garage door takes about 13 seconds to move from open to closed.)

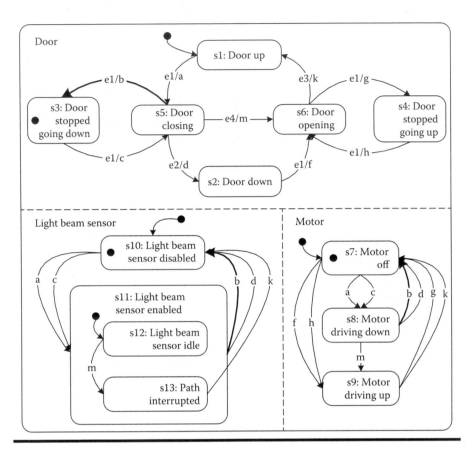

Figure 7.10   Event e1 causes broadcast output b to motor and light beam sensor.

In Figure 7.10, the second control signal event (e1) moves the system from s5: door closing to s3: door stopped going down. In the Motor orthogonal region, the broadcast event b causes the transition from state s8: motor driving down to state s7: motor off. In the Light Beam Sensor orthogonal region, broadcast event b causes a transition from s12: light beam sensor idle (and s11: light beam sensor enabled) to s10: light beam sensor disabled. At this point the system is event quiescent for an indefinite time interval.

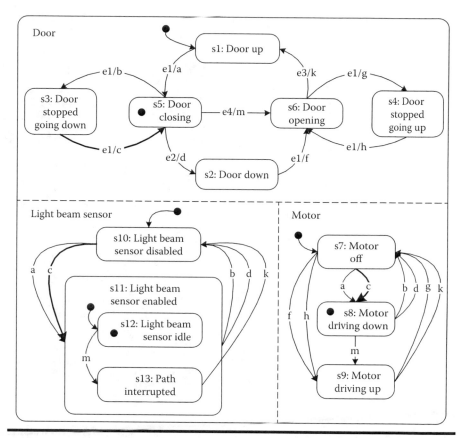

Figure 7.11   Event e1 causes broadcast output c to motor and light beam sensor.

In Figure 7.11, the third control signal event (e1) moves the system from s3: door stopped going down to s5: Door Closing and generates the broadcast event c. In the Motor orthogonal region, broadcast event c causes the transition from state s7: motor off to state s8: motor driving down. In the Light Beam Sensor orthogonal region, broadcast event c causes a transition from s10: light beam sensor disabled to s12: light beam sensor idle (and s11: light beam sensor enabled). At this point the system is temporarily event quiescent (as in Figure 7.9).

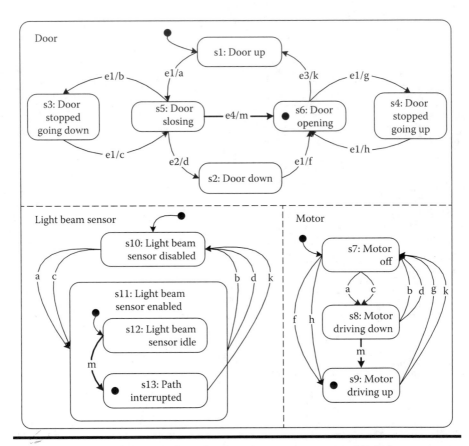

Figure 7.12 Event e4 causes broadcast output m to motor and light beam sensor.

In Figure 7.12, the Light beam interruption event (e4) moves the system from s5: Door Closing to s6: Door Opening and generates the broadcast event m. In the Motor orthogonal region, broadcast event m causes the transition from s8: motor driving down to s9: motor driving up. In the Light Beam Sensor orthogonal region, broadcast event m causes a transition from s12: light beam sensor idle to s13: path interrupted. At this point, the system is temporarily event quiescent (as in Figure 7.9) while the garage door is moving upward.

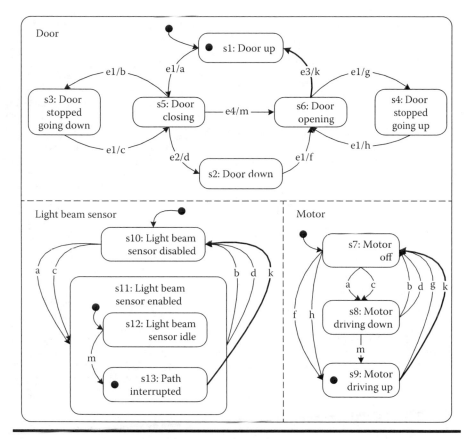

Figure 7.13   Event e3 causes broadcast output k to motor and light beam sensor.

In Figure 7.13, the end of up track event (e3) moves the system from s6: Door Opening to s1: Door Up and generates the broadcast event k. In the Motor orthogonal region, broadcast event k causes the transition from s5: motor driving up to s6: motor off. In the Light Beam Sensor orthogonal region, broadcast event k causes a transition from s11: path interrupted to s8: light beam sensor disabled. At this point the system is event quiescent for an indefinite time interval. All of this is summarized in the execution table for this scenario (Table 7.7).

The actual execution in the StateMate product breaks the steps in Table 7.7 into smaller, event specific steps to show the broadcasting more carefully. Table 7.8 shows this for the first two steps in Table 7.7.

Table 7.7 Execution Table for Figures 7.9 through 7.13

	Garage Door Statechart Execution Table			
	In States, before Event			*In States, after Broadcast*
Step	*Door, Motor, Light Beam Sensor*	*Input Event*	*Broadcast Output*	*Door, Motor, Light Beam Sensor*
0	s1, s7, s10	e1	a	s5, s8, s12
1	s5, s8, s12	e1	b	s3, s7, s10
2	s3, s7, s10	e1	c	s5, s8, s12
3	s5, s8, s12	e4	m	s6, s9, s13
4	s6, s9, s13	e3	k	s1, s7, s10

Table 7.8 StateMate-Style Execution Table for Figures 7.9 through 7.13

	Garage Door Statechart Execution Table			
Step	*In States, before Event*	*Input Event*	*Broadcast Output*	*In States, after Broadcast*
0.1	s1	e1	a	s5
0.2	s10	a		s12
0.3	s7	a		s8
1.1	s5	e1	b	s3
1.2	s12	b		s10
1.3	s8	b		s7

7.5 Lessons Learned from Experience

The StateMate product from the (now defunct) i-Logix corporation was introduced during the early years of my academic life. Therefore, I have no actual, industrial experience with the StateMate tool. In the early 1990s, I established a short-lived Case Tool Laboratory in Grand Rapids, Michigan. At one point, there were 25 commercial products in the lab, and StateMate was one of them. I took the i-Logix training, and had hands-on experience with the product for a week.

The main lesson I learned was that, except for very simple Statecharts, they are nearly impossible subjects for a technical inspection. The information is simply too dense. The i-Logix answer to this is that this is exactly why the Statechart engine is so important. Rather than hold a technical inspection of a Statechart, it is better to use the engine to execute various scenarios. Executing a Statechart then becomes very similar to the widely used technique of rapid prototyping for menu-driven applications—both are excellent for eliciting user/customer feedback before anything is even designed, nonetheless built. The second lesson is that limiting Statecharts to just the complex portions of a project is better than trying to model an entire, large project.

In the early 1990s, I attended a presentation that i-Logix representatives, including David Harel, gave to a Grand Rapids (Michigan) avionics company. They described their recent (and very impressive!) success at modeling a ballistic missile launch control system. After two weeks of working with the developers of the fully deployed system, they ran the Statechart model in the exhaustive mode. When the run was complete, the model had identified three sequences of events that would launch a missile.

At first, the developers scoffed—they knew of only two sequences of events that would launch a missile. After a close inspection, and much to their chagrin, the developers realized that the Statechart model revealed a third event sequence that would launch a missile. And this was in a fully deployed system!

7.6 Advantages and Limitations

Statecharts have two primary advantages: there are commercially available execution engines, and they scale up well to large, complex, and concurrent applications. The main disadvantage is complexity: both of the notation and the language on transitions. As a result, it is more difficult to perform a technical inspection of a complex Statechart. Table 7.9 is our continuing comparison of how models can express the nineteen behavioral issues.

Table 7.9 Representation of Behavioral Issues with Statecharts

Issue	Represented?	Example
Sequence	Yes	Sequential blobs
Selection	Yes	A blob with two emanating transitions
Repetition	Yes	A transition going back to a *previous* blob
Enable	Yes	The st(A) (Activity A is started) language element
Disable	Yes	The sp(A) (Activity A is stopped) language element
Trigger	Yes	The st!(A) (Activate the activity A) language element
Activate	Yes	The sp!(A) (Terminate the activity A) language element
Suspend	Yes	The sd!(A) (Suspend the activity A) language element
Resume	Yes	The rs!(A) (Resume the activity A) language element
Pause	Yes	(Suspend followed by resume)
Conflict	Yes	Determined from an execution table by the choice of which event occurs
Priority	Yes	Transition from preferred blob to other blob(s)
Mutual exclusion	Yes	Concurrent regions
Concurrent execution	Yes	Concurrent regions
Deadlock		Determined from an execution table
Context-sensitive input events	Yes	As in finite state machines
Multiple-context output events	Yes	As in finite state machines
Asynchronous events	Yes	Concurrent regions
Event quiescence	Yes	Determined from an execution table

Reference

[Harel 1988]
Harel, David, On visual formalisms. *Communications of the ACM* 1988; 31(5): 514–530.

Chapter 8

Swim Lane Event-Driven Petri Nets

For a long time, certainly since 1998, I have intuitively believed that Event-Driven Petri Nets (EDPNs, the topic of Chapter 6) are somehow equivalent in expressive power to David Harel's elegant Statecharts. One of the joys of teaching in a university graduate program is that, once in a while, a truly outstanding student comes along. This happened in the interval from 2010 to 2013 with Byron DeVries. In 2013, he presented his master's thesis, in which he proved that Swim Lane Marked Petri Nets are formally equivalent to UML Statechart types I, II, and III [DeVries 2013]. The germ of DeVries' thesis is that the Swim Lane concept of UML Activity Charts has a lot in common with the orthogonal regions of Statecharts. Using swim lanes to represent Statechart-concurrent regions was the missing piece of my intuitive belief. Extending DeVries' Swim Lane Marked Petri Nets to Swim Lane Event-Driven Petri Nets (SLEDPN) follows the extension of ordinary Petri nets to Event-Driven Petri Nets. That is to say, DeVries' Swim Lane Marked Petri Nets are a special case of the more general Swim Lane Event-Driven Petri Nets.

8.1 Definition and Notation

Definition [DeVries 2013]: A *swim lane marked Petri Net* is a seven-tuple (P, T, I, O, M, L, N) in which (P, T, I, O, M), is a marked Petri net and L is a set of n sets where

P is the set of places
T is the set of transitions

I is the input mapping of places in P to transitions in T

O is the output mapping of transitions in T to places in P

M is a marking that maps natural numbers to places in P

$n \geq 1$ is the number of swim lanes

L is the union of the places in the n lanes

N is the union of the transitions in the n lanes

Two easy secondary definitions follow almost directly:

Definition: A *swim lane (ordinary) Petri net* is a six-tuple (P, T, I, O, L, N) in which (P, T, I, O) is an ordinary Petri net (as in Chapter 5). The elements of the six-tuple are as in the first definition.

Definition: A *Swim Lane EDPN* is a seven-tuple (P, D, T, In, Out, L, N) in which (P, D, T, In, Out) is an Event-Driven Petri Net (EDPN) (as in Chapter 6). The elements of the seven-tuple are as in the first definition.

Symbolic notation for SLEDPN is shown in Figure 8.1.

8.1.1 Transition Enabling and Firing

Transition firing in a Swim Lane EDPN is exactly that for an EDPN. The same marking persistence conventions apply. The definitions from Chapter 6 are extended here.

Definition: A *marking M of a Swim Lane EDPN* (P, D, T, In, Out, L, N) is a sequence M = <m1, m2, ...> of p-tuples, where p is the sum of elements in the sets P and D, respectively. Individual entries in a p-tuple indicate the number of tokens in the event or data place.

Transition enabling and firing in a Swim Lane EDPN is exactly like that for regular EDPNs. Also, the definitions of events are the same, as well as the conventions for how long an output event remains marked.

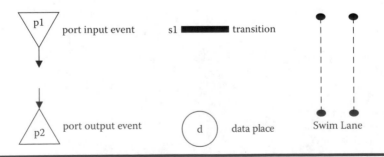

Figure 8.1 Swim Lane EDPN symbols.

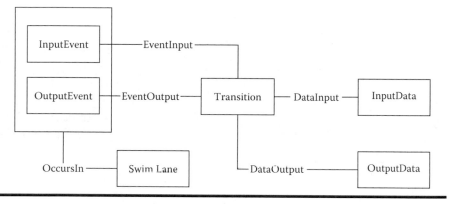

Figure 8.2   E/R model of a Swim Lane EDPN database.

8.1.2 Events in a Swim Lane Event-Driven Petri Net

Events in a Swim Lane EDPN are almost as they are in ordinary EDPNs. The definitions for both discrete and continuous events are the same, as well as the conventions for how long an output event remains marked. As the Swim Lanes can correspond to devices, many times it is convenient to associate port input and output events with specific swim lanes. As we have seen in Chapter 6, the E/R model in Figure 8.2 contains all the relations needed to fully describe a Swim Lane EDPN. All relations are many-to-many with optional participation, that is, their UML min/max descriptions are all (0...n).

The information in a Swim Lane EDPN database populated for a specific Swim Lane EDPN is both necessary and sufficient to recreate (except for spatial placement) a diagram of the Swim Lane EDPN.

8.2 Technique

8.2.1 Using Swim Lanes

Here are some general suggestions for using Swim Lane EDPNs:

1. Use transitions to represent actions.
2. Use places to represent any of the following:
   - Data
   - Pre- and postconditions
   - States
   - Messages
3. Use events to represent
   - Port input events (including passage of time)
   - Port output events
4. Use swim lanes to represent devices or concurrent processes.

5. Use the Input relationship to represent prerequisites and inputs to actions.
6. Use the Output relationship to represent consequences and outputs of actions.
7. Use markings to represent "states" of a net, memory, or counters.
8. Subsets of input data places to a transition can be used to define a context for input events.

Since ordinary Event Driven Petri Nets are a special case of Swim Lane EDPNs, the prompts described in Chapter 6 also apply to Swim Lane EDPNs as the following:

- Sequence
- Selection
- Repetition
- Enable, Disable, and Activate
- Trigger
- Suspend, Resume, and Pause
- Conflict and Priority
- Mutual exclusion
- Synchronization
- Context-Sensitive Input events
- Multiple Cause Output Events
- Event Quiescence

The best use of Swim Lane EDPNs is to closely examine the points at which devices interact. The ESML prompts presented in Chapter 5 are very handy for this purpose. This is illustrated in Figure 8.3, which depicts a portion of the Windshield Wiper example. Full Swim Lane EDPNs are usually very complex and space-consuming—good practice dictates focusing on specific areas of the problem being modeled (and tested). See Table 8.1 for more detailed information of the elements in the Swim Lane EDPN in Figure 8.3.

In Figure 8.3, the only states of the lever are Off and Intermittent, and the dial only has positions 1 and 2, corresponding to wiper speeds 6 and 12 strokes per minute. The lever transition from the Intermittent position to Off sends a trigger prompt to stop the wiper motor. The interactions between the Intermittent state and the dial positions are more complex—the lever and the dial each use an enable prompt. A wiper motor transition cannot fire until both enables are present. As the dial states are mutually exclusive, only one of the enable places can ever be marked at a time. Notice that, when a dial event occurs, the disable prompt removes a token from the enabling place, and places a token in the subsequent place. When the Int to Off transition fires, the tokens in both enable/disable places are consumed.

8.2.2 "Model Checking"

The quotation marks on the section heading are to differentiate the following discussion from the purely academic use of the term, which refers to heavily theoretical

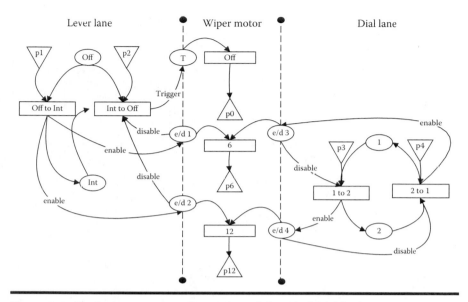

Figure 8.3 Swim Lane EDPN of a portion of the windshield wiper example.

Table 8.1 Legend for the Names in the Swim Lane EDPN Shown in Figure 8.3

Input Events	Output Events	Other Places
p1. Lever up one position	p0. Deliver 0 strokes per minute	Off: Lever at off
p2. Lever down one position	p6. Deliver 6 strokes per minute	Int: Lever at intermittent
p3. Dial left one position	p12. Deliver 12 strokes per minute	1. Dial at position 1
p4. Dial right one position		2. Dial at position 2
		T: a trigger place
		e/d: an enable/ disable place

techniques, along the lines of theorem proofs, to show that a model is correct. Much of that work centers on syntactic aspects of a model. Here we are more concerned with the underlying semantic meaning, because we wish to use a Swim Lane EDPN to generate test cases. Using scenarios is an excellent way to check whether a given Swim Lane EDPN represents our intent. Table 8.2 is an execution table for the Swim Lane EDPN description of just a portion of the Windshield Wiper Controller. In the scenario in Table 8.2, once the Wiper Lane transition 6 fires, place e/d-3 is unmarked, and the Dial Lane is deadlocked.

Table 8.2   Execution Table for the Swim Lane EDPN Shown in Figure 8.3

Step	Comment	Lever Lane		Wiper Lane		Dial Lane	
		Marked?	Enabled	Marked?	Enabled	Marked?	Enabled
0	Initial marking	Off				e/d-3, 1	
1	Lever move	p1, off	Off to Int enabled			e/d-3, 1	
2	Fire off to Int	Int, e/d-1, e/d-2			6	e/d-3, 1	
3	Fire 6	Int, e/d-2		p6		1	

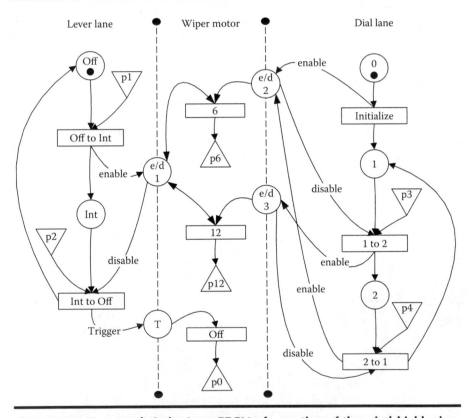

Figure 8.4   (Corrected) Swim Lane EDPN of a portion of the windshield wiper example.

I found this problem when I made the execution table (Table 8.2). I think the mechanism behind this goes back to two orthogonal views of any system—what it IS (a structural view) and what it DOES (a behavioral view). Executing a model with a set of scenarios can frequently reveal faults in the model being executed.

I was lucky, and found the deadlock fault on my first scenario. Figure 8.4 resolves this problem, and in the revision, describes a more accurate implementation. Some considerations:

- The lever and the dial are independent devices.
- The dial is only "active" when the lever is in the Intermittent position.
- There is no need for the two enable/disable places (e/d-1 and e/d-2) in Figure 8.3. The lever simply partially enables the wiper motor.
- There must be some way to initialize the dial to avoid the deadlock problem.

In Figure 8.4 there is just one enable/disable place (e/d-1) used by lever transitions for both wiper transitions 6 and 12. As these will always be mutually exclusive due to the mutual exclusive states of the dial, this is correct. When the Off to Int lever transition fires, the enable/disable place e/d-1 is marked. When the Int to Off transition fires, this place is unmarked. On the Dial side, the Initialize transition is added to "start" the dial. Figure 8.4 shows a likely initial marking—the Off and 0 places. Notice that the enable/disable place e/d-2 can be enabled by either the Initialize transition of by the 2 to 1 transition. There is a detailed part of the Wiper Motor swim lane. The transitions labeled "6" and "12" can only fire when both their respective enable/disable input places are marked. The double headed arrows to place e/d-1 assure that, once the wiper transitions are enabled, they remain enabled when a dial transition fires. The enable/disable place e/d-2 can be marked two ways—either by firing the Initialize transition or by firing the 2 to 1 transition. If transition 6 fires, place e/d-2 is unmarked. Place e/d-1 will remain marked. If transition 2 to 1 fires, it will mark place e/d-2 and the wiper transition 6 will be enabled again. If the Lever lane transition Int to Off fires (after a p2: lever down input), place e/d-1 will be unmarked, and the Wiper lane transition Off will be triggered. When it fires, whichever wiper motor (continuous) output was operating will cease. Table 8.3 is a very detailed marking sequence of these interactions.

8.2.3 Deriving Test Cases from a Swim Lane Event-Driven Petri Net

As noted earlier, the main value of a Swim Lane EDPN is that it enables a very detailed look at the interaction among swim lanes (usually devices). Clearly, such interactions should be tested, and the steps in an execution table lead directly to a sequence (not just a set) of very detailed test cases, as in Table 8.4. The CAUSE and VERIFY actions refer to system level events, and with a suitable harness, could be executed automatically. The "check memory" actions could be automated, given an appropriate tool. They might also have been checked at the unit testing level.

Table 8.3 Sample Execution Table for the Swim Lane EDPN shown in Figure 8.4

Step	Comment	Lever Lane		Wiper Lane		Dial Lane	
		Marked?	Enabled	Marked?	Enabled	Marked?	Enabled
0	Initial marking	Off	—	—	—	0	Initialize
1	Lever move	p1, off	Off to Int	—	—	0	Initialize
2	Fire off to Int	Int, e/d-1	—	—	—	0	Initialize
3	Fire initialize	Int, e/d-1	—	—	6	1, e/d-2	—
4	Fire 6	Int, e/d-1	—	p6	—	1	—
5	Event quiescent	Int, e/d-1	—	p6	—	1	—
6	Dial move	Int, e/d-1	—	p6	—	p3, 1	1 to 2
7	Fire 1 to 2	Int, e/d-1	—	p6	12	2, e/d-3	—
8	Fire 12	Int, e/d-1	—	p12	—	2	—
9	Event quiescent	Int, e/d-1	—	p12	—	2	—
10	Dial move	Int, e/d-1	—	p12	—	p4, 2	2 to 1
11	Fire 2 to 1	Int, e/d-1	—	p12	6	1, e/d-2	—
12	Fire 6	Int, e/d-1	—	p6	—	1	—
13	Event quiescent	Int, e/d-1	—	p6	—	1	—
14	Lever move	p2, Int, e/d-1	Int to Off	p6	—	1	—
15	Fire Int to off	Off, T	—	p6	Off	1	—
16	Fire off	Off	—	p0	—	1	—
17	Event quiescent	Off	—	p0	—	1	—

Table 8.4   Test Cases Identified from the Execution Table 8.3

Step	Comment	Test Objective	Action
0	Initial marking	Are the preconditions set correctly?	
1	Lever move	Does input event p1 operate?	CAUSE p1
2	Fire off to Int	Does the off to Int transition execute correctly?	check memory status of places Int and e/d-1
3	Fire initialize	Does the Initialize transition execute correctly?	check memory status of places 1 and e/d-2
4	Fire 6	Does p6 operate correctly?	VERIFY wiper speed = 6
5	Event quiescent	Does the wiper speed continue?	VERIFY wiper operation continues
6	Dial move	Does input event p3 operate?	CAUSE p3
7	Fire 1 to 2	Does the 1 to 2 transition execute correctly?	Check memory status of places 2, e/d-2 and e/d-3.
8	Fire 12	Does p12 operate correctly?	VERIFY wiper speed = 12
9	Event quiescent	Does the wiper speed continue?	VERIFY wiper operation continues
10	Dial move	Does input event p4 operate?	CAUSE p4
11	Fire 2 to 1	Does the 2 to 1 transition execute correctly?	Check memory status of places 1, 2, e/d-2 and e/d-3.
12	Fire 6	Is the wiper speed 6 strokes per minute?	VERIFY wiper speed = 6
13	Event quiescent	Does the wiper speed continue?	VERIFY wiper operation continues
14	Lever move	Does input event p2 operate?	CAUSE p2
15	Fire Int to off	Does the Int to off transition execute correctly?	check memory status of places Int, off, T, and e/d-1
16	Fire off	Does p0 operate correctly?	VERIFY wiper speed = 0
17	Event quiescent	Has the wiper motor stopped?	VERIFY wiper motor stopped

The test cases derived from the scenario described in Tables 8.3 and 8.4 cover the following structural (model-based) criteria:

- Every input event
- Every output event
- Every input place
- Every output place
- Every transition
- Edge
- Every device (Swim Lane)

The system lever portion of the tests in Table 8.3 is shown in a more compact form in Table 8.4.

8.3 The Continuing Problems

8.3.1 The Insurance Premium Problem

There is no reason to apply the power of SWEDPNs to the Insurance Premium problem. There are no events nor devices in the Insurance Premium problem so its Swim Lane EDPN is a single swim lane, the same as its ordinary Petri net (see Section 5.4.1). The same test cases from Chapter 5 could be derived from the Swim Lane EDPN formulation.

8.3.2 The Garage Door Controller

The input and output events and states for the Swim Lane EDPN formulation of the Garage Door Controller are identical to those in the EDPN version (see Chapter 6, Section 6.4.2). The EDPN formulation resolves the problems indentified in Chapter 5 for the ordinary Petri net formulation. Specifically, we assume that input events can be "created" at any time by the User. This obviates the need for an initial marking that includes every input event, as for ordinary Petri nets. The persistent marking of output events is ameliorated by two conventions:

1. For mutually exclusive output events, a marked output event is unmarked when one of its mutually exclusive events is marked by transition firing, or
2. A marked output event remains marked until the "next" transition is fired.

By adding swim lanes for the actual devices, we can examine (and test) more closely the way separate devices interact. We can also look at more obscure aspects of the Garage Door Controller, particularly possible failure modes of the light beam sensor. Eventually, the full Swim Lane EDPN for the Garage Door Controller will need the following elements:

Input Events	Output Events	States
p1: control signal	p5: start drive motor down	s1: Door Up
p2: end of down track hit	p6: start drive motor up	s2: Door Down
p3: end of up track hit	p7: stop drive motor	s3: Door Stopped Going Down
p4: laser beam crossed	p8: reverse motor down to up	s4: Door Stopped Going Up
		s5: Door Closing
		s6: Door Opening

8.3.2.1 The Light Beam Sensor

We begin with the connections among the Garage Door, the Control Device, the Light Beam Sensor, and the Motor. Recall that the Light Beam Sensor can only operate when the door is closing (moving down). We show this interaction with the enable/disable pattern and a trigger in Figure 8.5.

To follow the interaction across the swim lanes, suppose we have an initial marking of place s1: Door Up. This is a point of event quiescence. Next, suppose we cause a control signal, p1, which enables the Start Closing transition. When it fires, it sends a trigger prompt to the Start Motor Down transition, an enable prompt to the Sense Interruption transition, and marks place s5: Door Closing. We can assume that the Start Motor Down transition fires immediately after the trigger prompt, marking event p5: start drive motor down. Output event p5 will remain marked until another, mutually exclusive, motor event is marked. The system is now at another point of event quiescence. Two input events can occur at this point: either p1 or p4. We consider them separately.

Case 1: If event p1: control signal occurs, the Stop Closing transition is enabled. When it fires, the token in the enable/disable place is consumed, thereby disabling the Sense Interruption transition. It also consumes the token in place s5: Door Closing, and puts a token in place s3: Door Stopped Going Down. Finally, it sends a trigger prompt to the Stop Motor transition. When that transition fires, output event p7: stop drive motor is marked, and (due to mutual exclusion) event p5 is unmarked. At this point, the Swim Lane EDPN is deadlocked. No event can enable a transition.

Case 2: If event p4: laser beam crossed occurs, the Sense Interruption transition is enabled. When it fires, it consumes the token in the enable/disable place,

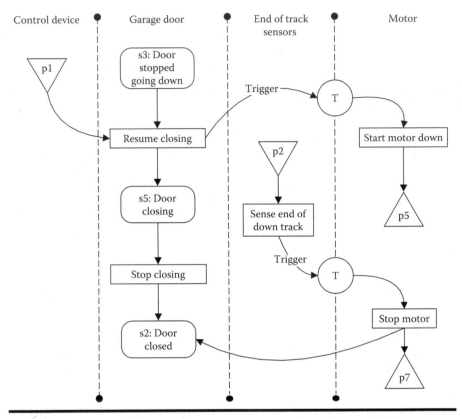

Figure 8.5 Swim Lane EDPN showing interaction among four swim lanes.

and sends a trigger prompt to the Reverse Down to Up transition. When it fires, the Reverse Down to Up transition unmarks place p5, and marks the mutually exclusive output event p8: reverse motor down to up. As, in this scenario case, the door is now opening, the Reverse Down to Up transition also marks place s6: Door Opening. As in Case 1, at this point, the Swim Lane EDPN is deadlocked. No event can enable a transition.

8.3.2.2 The End-of-Track Sensors

In a sense, Figure 8.6 commences from the end of Case 2, where the end states are s3: Door Stopped going Down, and s6: Door Opening.

The initial marking in Figure 8.6 is state s3: Door Stopped going Down. This is a point of event quiescence. Of the two input events in Figure 8.6, there is nothing that could physically cause event p2: end of down track. If event p1 occurs, the Resume Closing transition is the only transition that is enabled. When it fires, the trigger prompt fires the Start Motor Down transition, and event p5 occurs. When

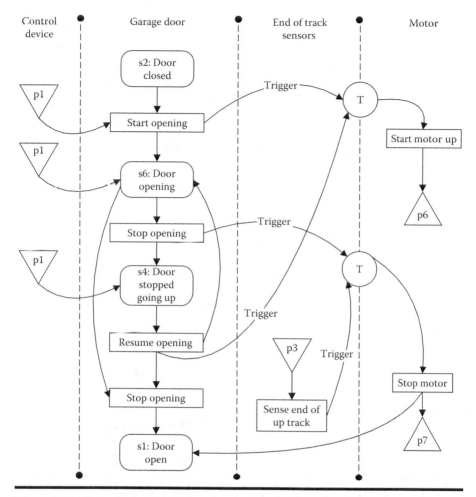

Figure 8.6   Swim Lane EDPN showing interaction with End-of-Track sensors.

the garage door reaches the end of the downward track, event p2: end of down track occurs. This enables the Sense End of Down Track transition, which, when it fires, triggers the Stop Motor transition. When the Stop Motor transition fires, the output event p7: stop drive motor occurs, and that unmarks output event p5, and also, state s2: Door Closed is marked.

8.3.2.3 The Door Opening Interactions

Figure 8.7 goes through the door opening sequence. The initial state is s2: Door Closed. This is a point of event quiescence. If input event p1 occurs, the Start Opening transition is enabled, and when it fires, it marks place s6: Door Opening, and sends the trigger prompt to the Start Motor Up transition. That transition fires

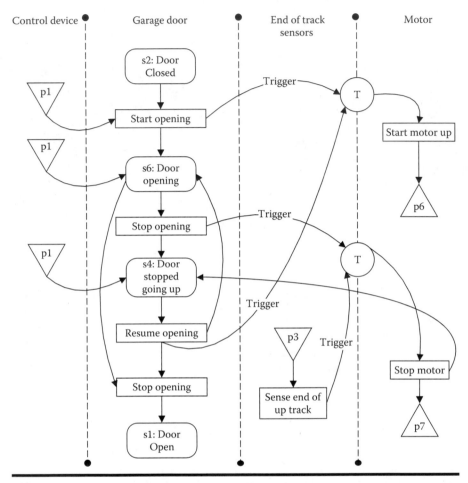

Figure 8.7   Swim Lane EDPN showing door opening options.

immediately, marking output event p6: start drive motor up, and the garage door is moving upward. As we saw with the door closing sequence (Figure 8.5) two input events can occur when state s6: Door Opening is marked—either p1: control signal, or p3: end of up track. The sequence that follows input event p3 is very similar to the events in Figure 8.6.

If event p1 occurs again, the Stop Opening transition fires, marking place s4: Door Stopped Going Up and sending a trigger prompt to the Stop Motor transition. When that transition fires, p6 is unmarked, p7 is marked, and place s4: Door Stopped Going Up is marked. Theoretically this start/stop loop could continue indefinitely. (In fact, there is a similar loop in the closing sequences.) In practice, pressing the control button is a manual operation that may take 10 milliseconds (similar to the time to touch a digit on a telephone digit keypad). The reaction time of the motor stopping is probably about a second, and in this time, the door has moved upward a measurable distance. The practical limit on the number of repetitions of the stop/resume cycle probably is about 20. Suffice it to say, this is not an infinite loop.

8.3.2.4 Failure Mode Event Analysis

Failure Mode Event Analysis deals with physical devices that can fail. Devices can fail for a variety of reasons—physical deterioration, excessive heat, voltage spikes, and so on. Regardless of the underlying cause, there are three failure modes, summarized in Table 8.5. Here we only consider the two common failure modes: Stuck-at-One (SA-1) and Stuck-at-Zero (SA-0) for the Light Beam Sensor. If the light beam sensor is stuck-at-zero and the physical event p4: light beam interruption occurs, no signal is sent. Symmetrically, if the light beam sensor is SA-1, the signal is always sent, even when the physical event p4 does not occur. We could probably consider the Intermittent failure mode by assigning probabilities to the SA-0 and SA-1 faults. It is important to remember (and model!) the fact that the physical input event can occur, but the device may fail. Figure 8.8 shows the normal operation of the light beam sensor.

Tables 8.6 and 8.7 are sample system level test cases that describe, respectively, the normal operation of door closing and normal light beam interruption. They use the CAUSE/VERIFY verbs first described in Chapter 1.

In normal operation, state s1: Door Open, is marked and the system is event quiescent. If input event p1: control device signal, occurs, the Start Closing transition is enabled. When it fires, it triggers the Start Motor Down transition and the door starts closing. The Start Closing transition also enables the Sense Interruption transition.

Table 8.5   Device Failure Modes

Stuck-at-Zero (SA-0)	Stuck-at-One (SA-1)	Intermittent
Does not send a signal when it should	Always sends a signal, even when it should not	Sometimes sends a signal when it should not; sometimes does not send a signal when it should, and usually cannot be repeated

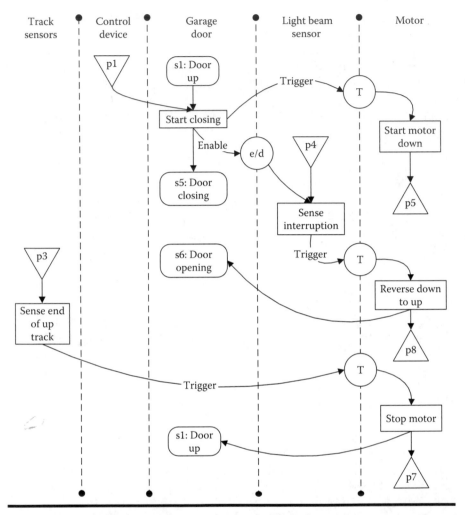

Figure 8.8   Normal operation of an interrupted light beam.

Case 1 (see Figure 8.7): If event p4: light beam interruption does not occur, the Sense Interruption cannot fire. Instead, the next event will be event 3. p2: end of down track hit. This enables the Sense End of Down Track transition which, when it fires, triggers the Stop Motor transition. The motor stops, and the garage door is closed.

Case 2 (see Figure 8.8): If event p4: light beam interruption occurs, the Sense Interruption is enabled and fires, sending a trigger to the Reverse Down to Up transition. When that transition fires (immediately), output event p8 is marked, and output event p5 is unmarked. When the door is fully open,

Table 8.6 System Level Test Case for Normal Door Closing.

Test Case	SysTC-1: Normal Door Closing			
Preconditions	1. Garage door is open			
Cause	**Occurs On**	**Verify**	**Occurs On**	**Observed Action**
1. p1: Control signal	Control device	2. p5: start drive motor down	Motor	Motor starts in down direction
				Door begins to close
3. p2: End of down track hit	End-of Track-sensors	4. p7: stop drive motor	Motor	Motor stopped
				Door is closed
Postconditions	1. Garage door is closed			

Table 8.7 System Level Test Case for Normal Door Closing with Light Beam Interruption

Test Case	SysTC-2: Normal Door Closing with Light Beam Interruption			
Preconditions	1. Garage door is open			
Cause	**Occurs On**	**Verify**	**Occurs On**	**Observed Action**
1. p1: Control signal	Control device	2. p5: Start drive motor down	Motor	Motor starts in down direction
				Door begins to close
3. p4: Laser beam crossed	Light beam sensor	4. p8: Reverse motor down to up	Motor	Motor reverses direction
				Door is opening
5. p3: End of up track hit	End-of-Track sensors	6. p7: Stop drive motor	Motor	Motor stopped
				Door is opening
Postconditions	1. Garage door is open			

input event p3: end of up track occurs, and the Sense End of Up Track transition fires, triggering the Stop Motor transition. When the Stop Motor transition fires, output event p7 is marked, event p8 is unmarked, and place s1: Door Up is marked. Table 8.4 is the corresponding system level test case. The Swim Lane EDPN model for the test case in Table 8.4 is shown in Figure 8.8.

Now, we consider the failure modes. We begin with the Stuck-at-Zero failure, described as a test case in Table 8.8, and a Swim Lane EDPN simulation in Figure 8.9.

Once test case SysTC-3 fails, a tester should try to determine the cause. The physical input event occurred, but the correct response did not occur. The SA-0 fault is a natural first choice. The Swim Lane EDPN in Figure 8.9 shows how the SA-0 fault could be simulated. Executing the Swim Lane EDPN in Figure 8.9 begins as the normal case in Figure 8.8. The difference is that the SA-0 place is not marked, and there is no transition that could possibly mark it. The Sense Interruption transition can still be enabled, and the p4: light beam interruption can still occur, but the transition can never fire. In Figure 8.9, the door will continue closing until the input event p2: occurs, and the Sense End

Table 8.8 System Level Test Case Result for Door Closing with SA-0 Light Beam Sensor Fault

Test Case	SysTC-3: Normal Door Closing with Light Beam Sensor SA-0 Fault			
Preconditions	1. Garage door is open			
	2. Light beam sensor has SA-0 fault			
Cause	**Occurs On**	**Verify**	**Occurs On**	**Observed Action**
1. p1: Control signal	Control device	2. p5: Start drive motor down	Motor	Motor starts in down direction
				Door begins to close
3. p4: Laser beam crossed	Light beam sensor	4. p8: Reverse motor down to up	Motor	Motor continues in down direction
At this point, the test case fails. Test case execution should stop.				

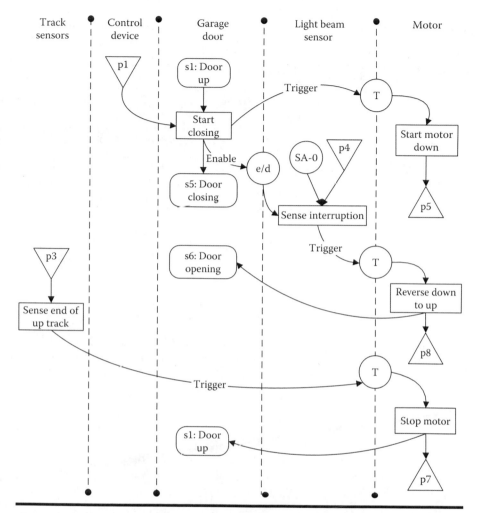

Figure 8.9   Simulating the SA-0 failure mode for the light beam sensor.

Of Down Track fires, triggering the Stop Motor transition. The end result is that the door is closed.

Table 8.9 describes the Stuck-at-1 fault, and it is simulated in Figure 8.10. As with the SA-0 fault, once the test fails, the tester should determine the cause. Of these two faults, the SA-0 fault could potentially cause harm or injury. The SA-1 fault only makes it impossible to automatically close the garage door.

Table 8.9 System Level Test Case Result for Door Closing with SA-1 Light Beam Sensor Fault

Test Case	SysTC-4: Normal Door Closing with Light Beam Sensor SA-1 Fault			
Preconditions	1. Garage door is open			
	2. Light beam sensor has SA-0 fault			
Cause	**Occurs On**	**Verify**	**Occurs On**	**Observed Action**
1. p1: control signal	Control device	2. p5: start drive motor down	Motor	Motor starts in down direction
				Door begins to close
				Motor reverses to up direction
				Motor stops, Door is open.
At this point, the test case fails. Test case execution should stop.				

The Stuck At 1 fault is simulated in Figure 8.10. A simple way to simulate the SA-1 fault is to just eliminate the input event p4: light beam interruption from the Swim Lane EDPN. It is replaced with the marked SA-1 place. As that place is both an input to and an output of the Sense Interruption transition, it will always be marked. Once event p1 occurs, the door starts closing, and immediately, it starts opening, until the end of the up track is reached.

Simply eliminating the p4 event seems like cheating. In fact the event may or may not occur, but regardless, the trigger should be sent to the Reverse Down to Up transition. A "nicer" way to show just this part is in Figure 8.11.

Notice the two connections from output event p4 in Figure 8.11. One (with the arrowhead) is the usual connection. The other, with the small circle termination, is an Inhibitor arc, defined next.

Definition: An *inhibitor arc* contributes to the enabling of a transition only when it is not marked.

In Figure 8.11, there are two transitions that can trigger the Reverse Down to Up transition. Assuming that the enable/disable place has already been marked

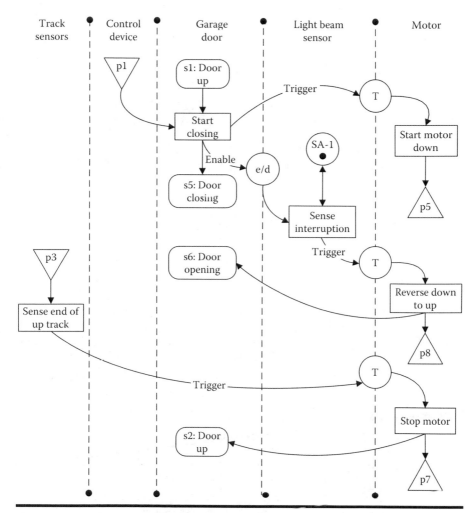

Figure 8.10 Simulating the SA-1 failure mode for the light beam sensor.

(by the door closing), the port input event p4: light beam interruption either does or does not occur. If it occurs, the Sense Interruption (with signal) is enabled, and when it fires, it triggers the Reverse Down to Up transition. If the p4 event does not occur, the Sense Interruption (no signal) transition is enabled due to the inhibitor arc connection. This representation is more accurate, because the output event p4: light beam interruption is a physical event in the real world. The mechanism that reacts to the inputs and causes the trigger is the item that is Stuck At 1.

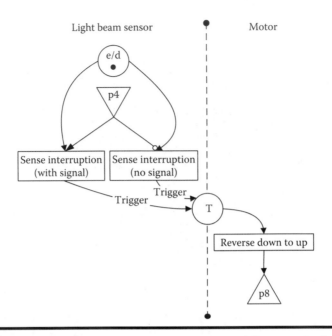

Figure 8.11 More accurate simulation of the SA-1 failure mode for the light beam sensor.

8.4 Deriving Test Cases from Swim Lane Event-Driven Petri Nets

Deriving test cases from a Swim Lane EDPN is very similar to the process for deriving test cases from an ordinary EDPN. The difference is that the devices corresponding to the swim lanes need to be added to a test case. Echoing the automatic test execution system is briefly described in Chapter 1, here is a short test case corresponding to the test case in Table 8.6. Reserved words are capitalized, and arguments must be selected from a predefined list (italic font). Noise words are permitted for readability (normal font).

Test Execution Script for test case SysTC-1
Preconditions: 1. Garage door is open
CAUSE the input event *p1: Control signal* ON the *Control Device*
VERIFY that the output event *p5: Start drive motor down* occurs ON the *Motor*
CAUSE the input event *p2: End of down track hit* ON the *End-of-Track Sensors*
VERIFY that the output event *p7: Stop drive motor* occurs ON the drive *Motor*
Postconditions: 1. Garage door is closed
Depending on the harness used with the automatic test executor, the preconditions could be **Caused** and the postconditions could be **Verified**.

8.5 Lessons Learned from Experience

I was part of a small (three people) team that built the test execution system indicated above. It was originally intended just for regression testing, but it was later extended, and became a commercial product with a lifetime of at least 15 years. We intended the noise words for readability, but the ever clever test developers soon wrote phrases such as

> If you are in a good mood, CAUSE the usual first input event *p1: control signal* ON oh, I dunno, maybe the *Control Device*.

Those of us who were slow keyboarders would simply write

> CAUSE *p1: Control signal* ON *Control Device*

Both versions would be interpreted into the same command sequence to the harness. At least our test case writers had a geeky sense of humor.

Other than that, SLEDPNs are so new; there is no actual direct experience. It is my hope that the MBT tool providers grow in this direction.

Table 8.10 is our continuing comparison of how models can express the 19 behavioral issues.

Table 8.10 Representation of Behavioral Issues with Swim Lane Event-Driven Petri Nets

Behavioral Issue	Represented in SLEDPNs?
Sequence	Yes
Selection	Yes, with inhibitor arcs
Repetition	Yes
Enable	Yes
Disable	Yes
Trigger	Yes
Activate	Yes
Suspend	Yes
Resume	Yes
Pause	Yes
Priority	Yes

(Continued)

Table 8.10 (*Continued*)   Representation of Behavioral Issues with Swim Lane Event-Driven Petri Nets

Behavioral Issue	Represented in SLEDPNs?
Mutual exclusion	Yes
Concurrent execution	Yes
Deadlock	Yes
Context-sensitive input events	Yes
Multiple-context output events	Yes
Asynchronous events	Yes
Event quiescence	Yes

Reference

[DeVries 2013]
DeVries, Byron, *Mapping of UML Diagrams to Extended Petri Nets for Formal Verification.* Master's Thesis, Grand Valley State University, Allendale, Michigan, USA, 2013.

Chapter 9

Object-Oriented Models

The Unified Modeling Language (UML) from the Object Management Group (for more details see http://www.omg.org/spec/UML/2.5) is the *de facto* standard for object modeling. In UML version 2.5 there are three categories of models—those that describe structure, behavior, and interaction. One of the main contributions of UML is the blending of the IS (structure) and DOES (behavior) views, as discussed in Chapter 1. The defining document, in PDF form, contains 748 pages. Given this size, this chapter is clearly just an introduction. The main goal here is to show how selected UML models can be (and are) used in commercial model-based testing (MBT) products. For the sake of completeness, the full list of UML 2.5 diagrams is listed here [OMG 2011]:

Structural Models:

- Class diagram
- Object diagram
- Component diagram
- Composite structure diagram
- Package diagram
- Deployment diagram

Behavioral Models:

- Use case diagram
- Activity diagram
- State machine diagram

Interaction Models:

- Sequence diagram
- Communication diagram
- Timing diagram
- Interaction overview diagram

The sheer size of UML 2.5 is almost an obstacle to its efficient use. For example, there are nine connectors (called "graphic paths") defined for structure diagrams. Seven of these are shown in Figure 9.1; Figures 9.2 through 9.4 show additional UML symbols for the various types of diagrams. In one sense, having such a rich symbolic vocabulary enables a very specific description. This is probably offset by both mnemonic difficulty at the user level, and tool support at the commercial level. There is a Pareto principle operational with the range of UML diagram symbols—users might use only 20% of the visual vocabulary.

Practitioners divide into two distinct schools of thought on using UML for software development. The top-down school begins with structural models (primarily

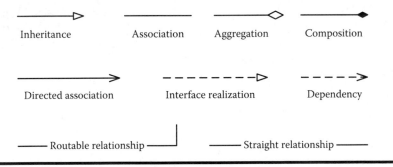

Figure 9.1 UML connector symbols.

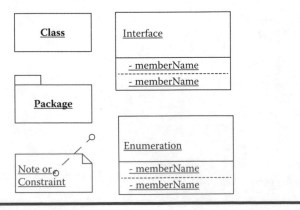

Figure 9.2 UML class symbols.

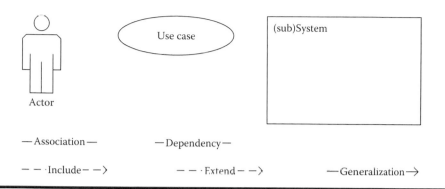

Figure 9.3   Use case symbols.

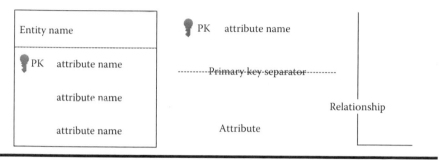

Figure 9.4   E/R and database symbols.

class diagrams) and then moves on to behavioral models (use case diagram, activity diagrams, and state machine diagrams). The bottom-up school starts with behavioral models, and uses them to help identify structural components. Both approaches work, and the choice is really one of personal style and taste.

9.1   Notation and Technique

In this chapter, we focus on the following requirements level, behavioral models:

- Use case diagram
- Activity diagram
- State machine diagram
- Sequence diagram

The use case diagram simply provides an overview of the connections among actors (sources and destinations of system level inputs and outputs) with individual use cases. In some ways, this is reminiscent of the Context Diagram of (Yourdon style)

Structured Analysis. Activity diagrams are a blend of ordinary Petri nets and flow-charts. As such, they are intuitively obvious, easy to create, and easily understood. The state machine diagram of choice is a simplified version of Statecharts (see Chapter 7). Finally, there is the sequence diagram, which is one of the few models anywhere that directly combines the IS and DOES views.

9.1.1 Use Case Diagrams

Use cases are an excellent way for customer/users and developers to communicate the DOES view of a system. There is a whole taxonomy of use cases, based generally on level of detail [Larman 1998]. Here is a sample use case for the Windshield Wiper. Once a set of use cases has been defined, the use case diagram shows how the individual use cases are related to external actors, which are sources of system level inputs and destinations of system level outputs. Figure 9.5 shows a sample use case diagram (with three other, not defined here, use cases) for the Windshield Wiper.

Use Case Name:	Lever up
Use Case ID:	L1
Initiating Actor(s):	Lever
Description:	Car driver moves the windshield lever up one position, for example, from the Off to the Intermittent position.
Preconditions:	1. An initial lever position. One of off, Int, or low.
Event Sequence:	1. Car driver moves the windshield lever up one position. 2.1. If the Dial position is 1, the system delivers 6 wipes per minute. 2.2. If the Dial position is 2, the system delivers 12 wipes per minute. 2.3. If the Dial position is 3, the system delivers 20 wipes per minute. 2.4. If the lever is at Low, the system delivers 30 wipes per minute. 2.5. If the lever is at High, the system delivers 60 wipes per minute.
Receiving Actor(s):	None
Postconditions:	1. Lever position is one of Int, low, or high
Source of Use Case:	Problem statement

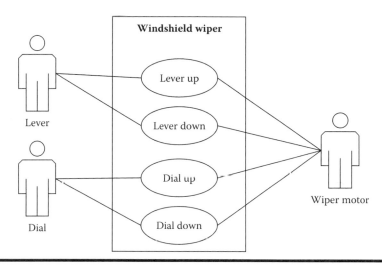

Figure 9.5 Use case diagram for the Windshield Wiper.

Development based on use case definition is strictly a bottom-up, behavior-based approach. How many use cases are necessary (or even sufficient) is an unavoidable question. The agile development answer is when the customer/user is satisfied with an existing set of use cases, but this may or may not meet the needs of developers and system testers. There are better answers, based on some notion of model coverage. Consider a set of following use cases that covers (choose a level or a combination of levels):

- All system level inputs
- All system level outputs
- All classes
- All messages
- All possible preconditions
- All possible postconditions

UML allows connections among use cases in a use case diagram. For example, one use case may use another, or a use case may extend (be a special case of) another. One problem with the use case diagram is that it does not scale up very well when many use cases are necessary.

9.1.2 Activity Diagrams

The UML activity diagram provides a closer look at the implementation of a single use case. The most interesting part of an activity diagram is the use of partitions, also called swim lanes (as adapted in Swim Lane EDPNs), to show the devices or

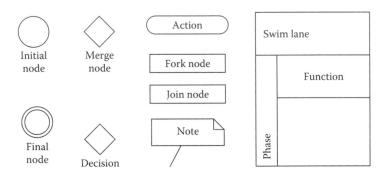

Figure 9.6 Symbols used in activity diagrams.

classes that perform activities. Figure 9.6 shows the symbols used in UML activity diagrams.

There are two conventions for activity diagrams: horizontal and vertical. In the horizontal version, an activity begins with an initial state in the upper left corner of the diagram, and generally proceeds from left to right, as in (western) reading style. The vertical style is more common when partitions (swim lanes) are used to designate either classes or agents that support the use case.

The symbol set suggests, correctly so, that activity diagrams are extensions of the basic idea of flowcharts to an object-oriented paradigm. The main difference is the use of the transition symbol, which acts almost like a Petri net transition. The fork and join possibilities refer to transitions having, respectively, several outputs and several inputs. More specifically, a forked transition corresponds exactly to the Petri net synchronized start, and a joined transition corresponds exactly to the Petri net synchronized stop (see Chapter 5). Figure 9.6 shows an activity chart for the Windshield Wiper use case described earlier.

Because activity diagrams are "object-oriented flowcharts," they inevitably have all the limitations of ordinary flowcharts. They do not express events well (in Figure 9.7, we use an action state to indicate that an event causes a change in lever position). The fact that they only show one sequence is not a limitation, because they are intentionally at the single use case level.

9.1.3 Statechart Diagrams

Statecharts, and their special, simplified case, finite state machines, are the control model central to UML 2.5. A Statechart is developed for each class, and therefore summarizes how a class participates in overall system behavior. One fundamental problem: there is no general way to compose Statecharts into larger Statecharts that represent behavior at the scope of two or more classes.

Several UML Statechart symbols (see Figure 9.8) are identical to those used in UML activity charts—this is deliberate. Personally, I do not see much use for

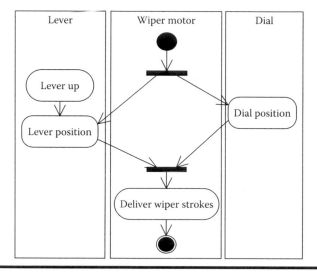

Figure 9.7 Activity diagram for sample use case.

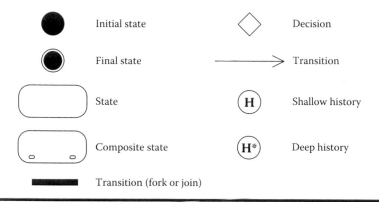

Figure 9.8 Symbols used in UML Statecharts.

the decision symbol in a Statechart. The state symbol is used for ordinary states, as in finite state machines. The composite state symbol is used when a state contains lower level states, as in the full Statechart notation. Technically, composite states can be used to describe concurrent regions, again as with the original Statechart notation. This is seldom used, however, because the UML assumption is that a particular class is being described. The shallow and deep history symbols are really directed toward tool support. A shallow history node contains the previous state of an object, and a deep history node contains the full history of an object.

9.1.4 Sequence Diagrams

The UML sequence diagram is unique in the world of mainline models: combines both the IS and the DOES views of a single use case. In a UML sequence diagram, an object from each class needed to implement the use case has an object lifeline, which refers to the execution time during which the object is instantiated. Lifelines are vertical dashed lines, with the name of the object/class at the top. The remaining symbols refer to the message-based communication among the objects (see Figure 9.9).

In a UML sequence diagram, we picture time as flowing downward. One variation of sequence diagrams is to enlarge the dashed line to a narrow rectangle to show when an object is instantiated and when it is destroyed. A sequence diagram refers to a single use case, it shows the message-based interaction among objects instantiated at execution time from the classes that support the use case. The objects/classes represent the IS view, and the sequence of messages and message returns captures the execution-time DOES view.

The sequence diagram in Figure 9.10 presumes the existence of four supporting classes—Lever, Controller, Dial, and Wiper. In Figure 9.10, the sequence begins

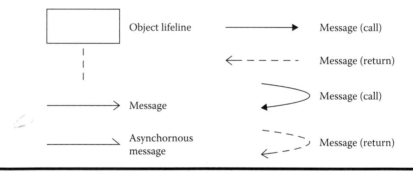

Figure 9.9   UML sequence diagram symbols.

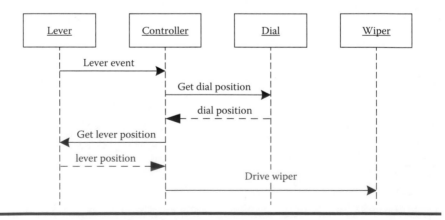

Figure 9.10   Sequence diagram for the sample use case.

with the lever event described as event 1 in the Event Sequence part of the use case given earlier. A "lever event" is somewhat vague, but that is the way it is expressed in the use case. We could have been very specific and describe the following several lever events:

- e1.1 Move lever from off to Int
- e1.2 Move lever from Int to low
- e1.3 Move lever from low to high
- e2.1 Move lever from high to low
- e2.2 Move lever from low to Int
- e2.3 Move lever from Int to off

This would yield six use cases, and each use case would need to have preconditions referring to the dial position. This issue of use case granularity echoes the discussion (see Chapter 3) of generalized events, which frequently are context-sensitive, versus a related list of context-specific events. At the requirements level, the generalized form reduces the number of use cases, and thereby enhances system comprehension. At the same time, however, a system tester would prefer more specific use cases.

9.2 Examples

Because the UML 2.5 models do not really add much to what we have already discussed more thoroughly in the preceding chapters, we will just comment on which models would be appropriate for the Windshield Wiper Controller.

- Figure 9.5 is the use case diagram
- Figure 9.7 is the activity diagram
- Figure 9.10 is the sequence diagram
- Figure 9.18 is the Statechart

9.3 The Continuing Problems

9.3.1 The Insurance Premium Problem

Figures 9.11 and 9.12 show two variations of detail in UML Activity Diagram for the Insurance Premium Problem. The Activity Diagram in Figure 9.11 is strictly linear, and is a simplification of the flowchart in Chapter 2. It prescribes the order in which inputs must be obtained and the order of decisions needed.

The diagram in Figure 9.12 uses more of the power of Activity Diagrams, specifically the fork and join points. The forks indicate that the two actions can be done in either order. The joins indicate that both actions must be complete before

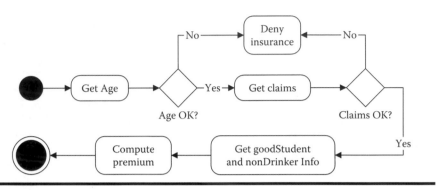

Figure 9.11   A linear activity diagram for the Insurance Premium Problem.

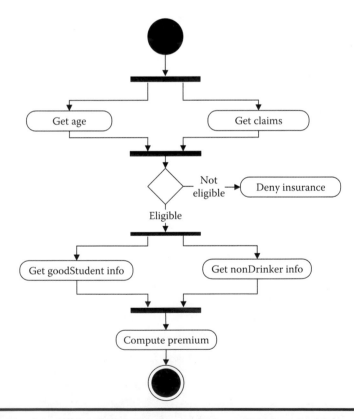

Figure 9.12   An activity diagram for the Insurance Premium Problem with Forks and Joins.

the next operation begins. We could still elaborate (belabor?) this further by postulating that the four data inputs come from four different organizations, each of which could have its own Swim Lane. The notation supports this, but adds very little to communication or comprehension.

The other UML diagrams considered in this chapter (use case diagram, Statechart, and sequence diagram) add very little to this problem.

9.3.2 The Garage Door Controller

9.3.2.1 Activity Diagram

As noted in the previous chapters, the Garage Door Controller is an event-driven system. As such, it is not well served by the Activity Diagram in Figure 9.13. Activity diagrams, particularly with swim lanes, consume a lot of space on a drawing. There was no room for a fifth swim lane for the safety device (laser beam). Also, there is no easy way to show events—they appear as activity diagram actions; the same holds true for what would normally be understood as states. Figure 9.13 does

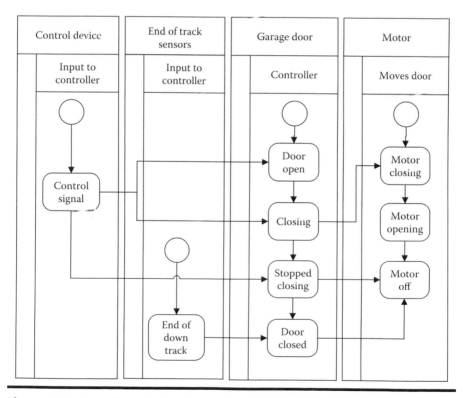

Figure 9.13   A partial activity diagram for the Garage Door Controller.

show all potential points at which the control device signal is used, but there is no sense of sequence. The output (motor) events, because they are in their own swim lane, can be interpreted as being truly concurrent with either of the input events in their swim lanes.

The Activity Diagram in Figure 9.13 is very incomplete—it has no swim lane for the laser beam safety device, and it only shows most of the door closing actions—the door opening actions are also missing. One conclusion: activity diagrams are probably best suited for unit level, computational applications.

9.3.2.2 Use Cases for the Garage Door Controller

Development (and modeling) with use cases is a bottom-up process. The attendant question is how does the modeler know when a set of use cases is sufficient? In this subsection, we provide two ways to answer this question—derivation of a finite state machine from the use cases, and use of incidence matrices relating use cases to input events and output actions. The Garage Door Controller example supports a useful discussion of this persistent question. For convenience, the input events, output actions, and states of an eventual finite state machine are repeated here. Although slightly renamed, they align exactly with the discussion of Chapter 4 (Section 4.4.2). In the development of use cases for the Garage Door Controller, we will use Larmann's "expanded essential use cases [Larmann 1998]."

Input Events	Output Events (Actions)	States
e1: Control device signal	a1: Start drive motor down	s1: Door is open
e2: End of down track reached	a2: Start drive motor up	s2: Door is closed
e3: End of up track reached	a3: Stop drive motor	s3: Door stopped going down
e4: Light beam interruption sensed	a4: Reverse motor down to up	s4: Door stopped going up
		s5: Door is closing
		s6: Door is opening

9.3.2.2.1 Expanded Essential Use Cases (First Try)

Since use case development is a bottom-up process, there is no guarantee that an initial set of use cases is a good set. We shall see some ways to determine if a set of use cases is a "good" set.

Name, ID	*Normal Open*		*EEUC-1*
Description	A fully closed door is opened with a control device signal		
Preconditions	s2: Door is closed		—
—	Event Sequence		
—	Inputs		Outputs
—	1. e1: Control device signal		2. a2: Start drive motor up
—	3. e3: End of up track reached		4. a3: Stop drive motor
Postconditions	s1: Door is open		—
Events used	e1, e3		a2, a3

Name, ID	*Normal Close*		*EEUC-2*
Description	A fully open door is closed with a control device signal		
Preconditions	s1: Door is open		—
—	Event Sequence		
—	Inputs		Outputs
—	1. e1: Control device signal		2. a1: Start drive motor down
—	3. e2: End of down track reached		4. a3: Stop drive motor
Postconditions	s2: Door is closed		—
Events used	e1, e2		a1, a3

Name, ID	*Door Stopped while Closing*	EEUC-3
Description	While door is closing, a control device signal occurs. The door stops.	
Preconditions	s5: Door is Closing	—
—	Event Sequence	
—	Inputs	Outputs
—	1. e1: Control device signal	2. a3: Stop drive motor
Postconditions	s3: Door is stopped going down	—
Events used	e1	a3

Name, ID	*Door Continues after Being Stopped while Closing*	EEUC-4
Description	After being stopped while closing, a control device signal occurs. The door continues closing.	
Preconditions	s3: Door is stopped going down	—
—	Event Sequence	
—	Inputs	Outputs
—	1. e1: Control device signal	2. a1: Start drive motor down
Postconditions	s5: Door is closing	—
Events used	e1	a1

Name, ID	*Door Stopped while Opening*	EEUC-5
Description	While door is opening, a control device signal occurs. The door stops.	
Preconditions	s6: Door is opening	—
—	Event Sequence	
—	Inputs	Outputs
—	1. e1: Control device signal	2. a3: Stop drive motor
Postconditions	s4: Door is stopped going up	—
Events used	e1:	a3

Name, ID	Door Continues after Being Stopped while Opening	EEUC-6
Description	After being stopped while opening, a control device signal occurs. The door continues opening	
Preconditions	s4: Door is stopped going up	—
—	Event Sequence	
—	Inputs	Outputs
—	1. e1: Control device signal	2. a2: Start drive motor up
Postconditions	s6: Door is Opening	—
Events used	e1	a2

Name, ID	Light Beam Stop	EEUC-7
Description	The light beam is crossed while door is closing. The door stops and then reverses direction.	
Preconditions	s5: Door is closing	—
—	Event Sequence	
—	Inputs	Outputs
—	1. e4: Light beam interruption sensed	2. a4: reverse motor down to up
Postconditions	s6: Door is Opening	—
Events used	e4	a4

9.3.2.2.1.1 Use Case/Event Incidence (First Try)

At a minimum, we need a set of use cases that "covers" the input events and output actions; the incidence matrix in Table 9.1 shows that the first try set of use cases satisfies this criterion.

Table 9.1 Incidence Matrix for the *First Try* Use Cases

EEUC-	e1	e2	e3	e4		a1	a2	a3	a4
1	x		x				x	x	
2	x	x				x		x	
3	x							x	
4	x					x			
5	x							x	
6	x					x			
7				x					X

9.3.2.2.1.2 Directed Graph of EEUCs (First Try)

The narrative descriptions of the first try use cases are repeated here. Table 9.2 contains the pre- and postconditions of these use cases.

> EEUC-1: A fully closed door is opened with a control device signal.
> EEUC-2: A fully open door is closed with a control device signal.
> EEUC-3: While door is closing, a control device signal occurs. The door stops.
> EEUC-4: After being stopped while closing, a control device signal occurs. The door continues closing.
> EEUC-5: While door is opening, a control device signal occurs. The door stops.
> EEUC-6: After being stopped while opening, a control device signal occurs. The door continues opening.
> EEUC-7: The light beam is crossed while door is closing. The door stops and then reverses direction.

To move toward a finite state machine that "contains" a set of use cases, we observe that, when preconditions of one use case align with postconditions of a second use case, the second use case can "connect" with the first one. This leads to the directed graph in Figure 9.14.

It seems a little awkward that EEUCs 1 and 2 are isolated from the other first try use cases. From a testing standpoint, it would be nice if we could use the use cases as "building blocks" that could be connected to form a wide variety of use cases. Consider this User Story (a long use case):

"A fully open door is closed with a control device signal. While it is closing, the light beam sensor is tripped. The door stops, and immediately starts to open. The

Table 9.2   Pre- and Postconditions of the *First Try* Use Cases

EEUC-	Precondition	Postcondition
1	s2: Door is closed	s1: Door is open
2	s1: Door is open	s2: Door is closed
3	s5: Door is closing	s3: Door is stopped going down
4	s3: Door is stopped going down	s5: Door is closing
5	s6: Door is opening	s4: Door is stopped going up
6	s4: Door is stopped going up	s6: Door is opening
7	s5: Door is closing	s6: Door is opening

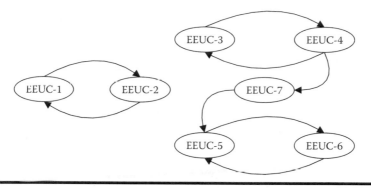

Figure 9.14   Directed graph of the first try EEUCs.

user touches a control device, and the door stops on the way up. After a second control device touch, the door continues opening. Eventually, the end of the up track is reached, and the door stops in the fully open position."

We cannot quite make this happen with the first try EEUCs, because there is no way to connect EEUC-2 to EEUC-7. This leads to the second try of EEUCs.

9.3.2.2.2 Expanded Essential Use Cases (Second Try)

To obtain the needed pre- and postconditions, we rewrite EEUC-1 as the pair EEUC-1a and EEUC-1b, and similarly, EEUC-2 is rewritten as use cases EEUC-2a and EEUC-2b. The other first five try use cases need no changes, so they are not repeated here.

Name, ID	Normal Open		EEUC-1a
Description	A fully closed door begins to open with a control device signal		
Preconditions	s2: Door is Closed		—
—	Event Sequence		
—	Inputs		Outputs
—	1. e1: control device signal		2. a2: start drive motor up
Postconditions	s6: Door is Opening		—
Events Used	e1		a2

Name, ID	Normal Open		EEUC-1b
Description	An opening door reaches the end of the up track.		
Preconditions	s6: Door is Opening		—
—	Event Sequence		
—	Inputs		Outputs
—	1. e3: end of up track reached		2. a3: stop drive motor
Postconditions	s1: Door is Open		—
Events Used	e3		a3

Name, ID	Normal Close		EEUC-2a
Description	A fully open door begins to close with a control device signal		
Preconditions	s1: Door is Open		—
—	Event Sequence		
—	Inputs		Outputs
—	1. e1: control device signal		2. a1: start drive motor down
Postconditions	s5: Door is Closing		—
Events Used	e1		a1

Name, ID	Normal Close		EEUC-2b
Description	A closing door reaches the end of the down track.		
Preconditions	s5: Door is Closing		—
—	Event Sequence		
—	Inputs		Outputs
—	1. e2: end of down track reached		2. a3: stop drive motor
Postconditions	s2: Door is closed		—
Events Used	e2		a3

9.3.2.2.2.1 Use Case/Event Incidence (Second Try)

As before, we need a set of use cases that "covers" the input events and output actions; the incidence matrix in Table 9.3 shows that the second try use cases satisfy this criterion.

The narrative descriptions of the second try use cases are repeated here. Table 9.4 contains the pre- and postconditions of these use cases.

Table 9.3   Incidence Matrix for the *Second Try* Use Cases

EEUC-	e1	e2	e3	e4		a1	a2	a3	a4
1a	x						x		
1b			x					X	
2a	x					x			
2b		x						X	
3	x							X	
4	x					x			
5	x							X	
6	x						x		
7				x					X

Table 9.4 Pre- and Postconditions of the *Second Try* Use Cases

EEUC-	Precondition	Postcondition
1a	s2: Door is closed	s6: Door is opening
1b	s6: Door is opening	s1: Door is open
2a	s1: Door is open	s5: Door is closing
2b	s5: Door is closing	s2: Door is closed
3	s5: Door is closing	s3: Door is stopped going down
4	s3: Door is stopped going down	s5: Door is closing
5	s6: Door is opening	s4: Door is stopped going up
6	s4: Door is stopped going up	s6: Door is opening
7	s5: Door is closing	s6: Door is opening

EEUC-1a: A fully closed door begins to open with a control device signal.

EEUC-1b: An opening door reaches the end of the up track.

EEUC-2a: A fully open door begins to close with a control device signal.

EEUC-2b: A closing door reaches the end of the down track.

EEUC-3: While door is closing, a control device signal occurs. The door stops.

EEUC-4: After being stopped while closing, a control device signal occurs. The door continues closing.

EEUC-5: While door is opening, a control device signal occurs. The door stops.

EEUC-6: After being stopped while opening, a control device signal occurs. The door continues opening.

EEUC-7: The light beam is crossed while door is closing. The door stops and then reverses direction.

The user story that could not be represented by a chain of first try use cases can now be represented as follows:

"A fully open door is closed with a control device signal (EEUC-2a). While it is closing, the light beam sensor is tripped. The door stops, and immediately starts to open (EEUC-7). The user touches a control device, and the door stops on the way up (EEUC-5). After a second control device touch, the door continues opening (EEUC-6). Eventually, the end of the up track is reached, and the door stops in the fully open position (EEUC-1b)."

9.3.2.2.2.2 Directed Graph of EEUCs (Second Try)

Figure 9.15 is a directed graph of the second try EEUCs. Notice we no longer have isolated use cases.

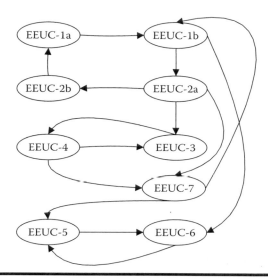

Figure 9.15   Directed graph of the second try EEUCs.

9.3.2.2.3 Long Expanded Essential Use Cases—(Possibly for Regression Testing [Third Version])

Some authors prefer "end-to-end" use cases, as seen next. Note that if these are tested successfully, several aspects of the problem had to work. This makes them good candidates for regression testing. These use cases also exercise event e1 in several contexts. Our discussion continues with the same input events, output events, and states as before.

Name, ID	Normal Open	LongFFUC-1
Description	A fully closed door is opened with a control device signal	
Preconditions	s2: Door is closed	—
—	Event sequence	
—	Inputs	Outputs
—	1. e1: Control device signal	2. a2: Start drive motor up
—	3. e3: End of up track reached	4. a3: Stop drive motor
Postconditions	s1: Door is open	—
Events used	e1, e3	a2, a3

Name, ID	*Normal Close*		*LongEEUC-2*
Description	A fully open door is closed with a control device signal		
Preconditions	s1: Door is open		—
—	Event Sequence		
—	Inputs		Outputs
—	1. e1: Control device signal		2. a1: Start drive motor down
—	3. e2: End of down track reached		4. a3: Stop drive motor
Postconditions	s2: Door is closed		—
Events used	e1, e2		a1, a3

Name, ID	*Open with Intermediate Stop*		*LongEEUC-3*
Description	A fully closed door is opened with a control device signal, then stopped, then started again.		
Preconditions	s2: Door is closed		—
—	Event Sequence		
—	Inputs		Outputs
—	1. e1: Control device signal		2. a2: Start drive motor up
—	3. e1: Control device signal		4. a3: Stop drive motor
—	5. e1: Control device signal		6. a2: Start drive motor up
—	7. e3: End of up track reached		8. a3: Stop drive motor
Postconditions	s1: Door is open		—
Events used	e1, e3		a2, a3

Name, ID	*Close with Intermediate Stop*		*LongEEUC-4*
Description	A fully open door is closed with a control device signal, then stopped, then started again.		
Preconditions	s1: Door is open		—
—	Event Sequence		
—	Inputs		Outputs
—	1. e1: Control device signal		2. a1: Start drive motor down
—	3. e1: Control device signal		4. a3: Stop drive motor

—	5. e1: Control device signal	6. a1: Start drive motor down
—	e2: End of down track reached	8. a3: Stop drive motor
Postconditions	s2: Door is closed	—
Events used	e1, e2	a1, a3

Name, ID	*Close with Light Beam Sensor Stop*	*LongEEUC-5*
Description	A fully open door is closed with a control device signal, and then reversed when the light beam is interrupted.	
Preconditions	s1: Door is open	—
—	Event Sequence	
—	Inputs	Outputs
—	1. e1: Control device signal	2. a1: Start drive motor down
—	3. e4: Light beam interruption sensed	4. a4: Reverse motor down to up
—	5. e3: End of up track reached	6. a3: Stop drive motor
Postconditions	s1: Door is open	—
Events used	e1, e3, e4	a1, a3, a4

Use Case/Event Incidence (Third Version)

With the long use cases, the input events and output actions are all covered, but concatenation is sacrificed. Long use cases are better suited for regression tests (Table 9.5).

9.3.2.3 Use Case Diagram for the Garage Door Controller

Figure 9.16 is the use case diagram for the second try (short) use cases. It could be derived from the use case/event incidence matrix. The icons, technically known as actors, are the sources of system level inputs and destinations of system outputs. Although they appear to be human figures, they can be any device. It is like the Context Diagram of the Yourdon-style dataflow diagrams. It does not provide much information, other than a quick visual overview of the system being modeled. Also, imagine a system that requires 30 or more use cases and many actors. The complexity of associations between actors and use cases would be obscured by the confusion of connecting lines.

Table 9.5 Incidence Matrix for the Long Use Cases

LongEEUC-	e1	e2	e3	e4		a1	a2	a3	a4
1	x		x				x	x	
2	x	x				x		x	
3	x		x				x	x	
4	x	x				x		x	
5	x		x	x		x		x	x

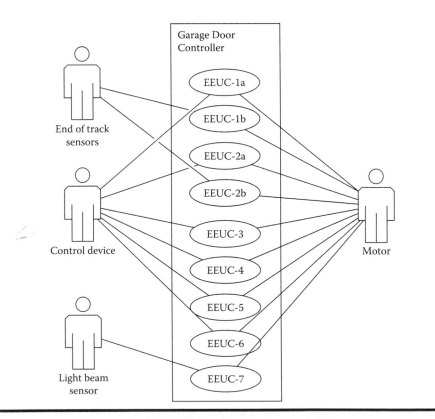

Figure 9.16 Use case diagram for the Garage Door Controller.

9.3.2.4 Sequence Diagram for the Garage Door Controller

Sequence diagrams represent the interaction of a single use case with the classes (or procedural units) that are needed to support its execution. This is the diagram that merges the "IS view" (structure) with the "DOES view" (behavior). In a sequence

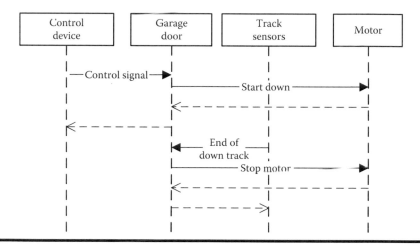

Figure 9.17 Sequence diagram for the Normal Close long use case.

diagram, the communication flow among classes is portrayed, and the sense of time is that it increases downward. The sequence diagram in Figure 9.17 is for the long use case LongEEUC-2 (Normal close). The messages from the devices are indicated by solid, named arrows; the returns are dashed arrows. The Garage Door class gets messages from the devices and controls the motor.

9.3.2.5 Statechart for the Garage Door Controller

Figure 7.8 in Chapter 7 is repeated here, and renamed as Figure 9.18. Only three orthogonal regions are shown in order to keep the door orthogonal region as close as possible to the finite state machine in Chapter 4. The input events from the control device and the track sensors are just shown as events that cause transitions. This makes the broadcasting feature of Statecharts easier to follow.

9.4 Deriving Test Cases from UML Models

9.4.1 Test Cases from Activity Diagrams

The UML Activity Diagrams are used by several commercial MBT products for test case generation. Generally speaking, they are used for computational applications, as exemplified by the Insurance Premium Problem. Due to their similarity to ordinary flowcharts, all of the observations made in Chapter 2 on test cases from flowcharts apply to deriving test cases from activity diagrams. The strong sequential

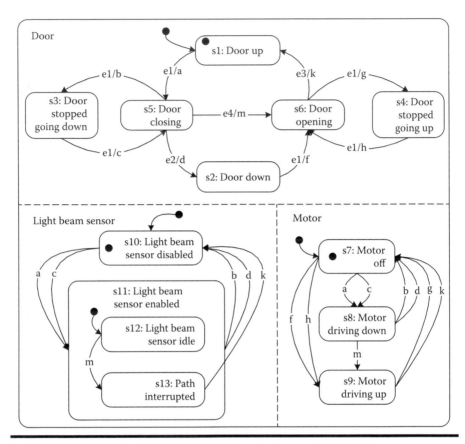

Figure 9.18 Statechart for the Garage Door Controller.

nature of activity charts translates, sometimes directly, into abstract test cases. Here is a quick summary:

- Paths in an activity diagram correspond to abstract test cases.
- The Fork and Join capabilities allow for generation of portions of a test case in different orders.
- Events are not easily shown.
- Context-sensitive events must be recognized manually.
- Event-driven applications are not well served by activity diagrams.

The vendor-provided examples in Chapters 14, 16, and 18 are particularly instructive.

9.4.2 Test Cases from Use Cases

An expanded essential use case is almost directly usable as a test case. Table 9.6 shows the similarities.

9.4.3 Test Cases from a Use Case Diagram

The UML Use Case Diagram offers no support for test case generation. At most, it may guide the order in which tests are grouped, presumably by actors.

9.4.4 Test Cases from Sequence Diagrams

UML Sequence Diagrams are clearly redundant with the use cases they represent. They could be used for a skeleton of an integration-level test case because they highlight the sources and destinations of internal communication.

9.4.5 Test Cases from Statecharts

As discussed in Chapter 7, Statecharts are excellent for test case generation, particularly for event-driven systems, and systems with independent devices. A path in a Statechart, across orthogonal regions if necessary, is a test case. One limitation is

Table 9.6   Comparison of Expanded Essential Use Cases and Test Cases

Expanded Essential Use Case Element	Test Case Content
Short name, descriptive name	Short name, descriptive name
Narrative description	Narrative description; business rule
Preconditions	Set-up conditions
Input event sequence	Input event sequence
Output event (action) sequence	Expected output event (action) sequence
	Observed output event (action) sequence
Postconditions	Resulting conditions
	Pass/fail result
	Date run
	System/software version

that input events are seen as causes of transitions; another is that output events are not generally shown—they are "indicated" by states (in device-dependent orthogonal regions) that describe when output events occur. In the presence of a Statechart engine, a user-driven sequence of input events can be used to generate a nearly complete system level test case.

9.5 Advantages and Limitation

As UML is so widely used, it has become a *de facto* standard, and many commercial MBT products offer some support for UML descriptions. Table 9.7 shows the behavioral issues that are represented by UML Activitiy Diagrams.

As UML uses the Statechart notation, we would expect that the representation of behavioral issues would be identical to that of Statecharts in Chapter 9.

Table 9.7 Representation of Behavioral Issues with UML Activity Diagrams

Issue	Represented?	Comment
Sequence	Yes	Main point of flowcharts
Selection	Yes	Main point of flowcharts
Repetition	Yes	Main point of flowcharts
Enable	No	Must describe as text in a activity box
Disable	No	Must describe as text in a activity box
Trigger	No	Must describe as text in a activity box
Activate	No	Must describe as text in a activity box
Suspend	No	Must describe as text in a activity box
Resume	No	Must describe as text in a activity box
Pause	No	Must describe as text in a activity box

(Continued)

Table 9.7 (*Continued*)   Representation of Behavioral Issues with UML Activity Diagrams

Issue	Represented?	Comment
Conflict	No	
Priority	No	Must describe as text in a process box
Mutual exclusion	No	Parallel paths after a decision
Concurrent execution	Partial	Use of the fork and join constructs
Deadlock	No	
Context-sensitive input events	No	Must be deduced by careful examination of sequences following a decision.
Multiple-context output events	Indirectly	Must be deduced by careful examination of sequences following a decision
Asynchronous events	No	
Event quiescence	No	
Memory?	Yes	By reference to previous decisions, inputs, and activity boxes
Hierarchy?	Yes	Activity boxes and be expanded into more detail as needed

This is not quite the case: several of the ESML issues are implemented in the elaborate transition language of the original Statecharts; but they are not part of UML 2.5.

Table 9.8 shows the behavioral issues that are represented by UML Statecharts.

Table 9.8 Representation of Behavioral Issues with UML Statecharts

Issue	Represented?	Example
Sequence	Yes	Sequential blobs
Selection	Yes	A blob with two emanating transitions
Repetition	Yes	A transition going back to a "previous" blob
Enable	No	
Disable	No	
Trigger	No	
Activate	No	
Suspend	No	
Resume	No	
Pause	No	
Conflict	Yes	Determined from an execution table by the choice of which event occurs
Priority	Yes	Transition from preferred blob to other blob(s)
Mutual exclusion	Yes	Concurrent regions
Concurrent execution	Yes	Concurrent regions
Deadlock		Determined from an execution table
Context-sensitive input events	Yes	As in finite state machines
Multiple-context output events	Yes	As in finite state machines
Asynchronous events	Yes	Concurrent regions
Event quiescence	Yes	Determined from an execution table

References

[OMG 2011]
OMG Unified Modeling Language™ (OMG UML) Superstructure, version 2.4.1, downloaded from http://www.omg.org/spec/UML/2.5.
[Larman 1998]
Larman, C., *Applying UML and Patterns: An Introduction to Object-Oriented Analysis and Design*. Prentice-Hall, Upper Saddle River, NJ, 1998.

Chapter 10

Business Process Modeling and Notation

The Business Process Modeling and Notation (BPMN) was developed by the Business Process Modeling Institute (http://www.bpminstitute.org/) and is currently managed by the Object Management Group. BPMN 2.0 (the present version) is intended to help organizations model and communicate their business process, both internal and with external entities. As in Chapter 9 on Unified Modeling Language (UML), the BPMN defining document, in PDF form, contains 538 pages. Given this size, this chapter is clearly just an introduction. The main goal here is to show how a BPMN model can describe a software development process that involves developers and management. This chapter treats BPMN for two reasons—a few commercial MBT tools support the notation (see Figure 14.1 from the Smartesting product and Figures 18.2 and 18.3 from sepp.med). Also, it appears in the ISTQB MBT Extension to the Foundation Level Syllabus.

10.1 Definition and Notation

The BPMN notation has four basic shapes: activities, gateways, events, and flows (see Figure 10.1). Each of these has several variations, resulting in a very comprehensive set of meaningful shapes (Figures 10.2 through 10.5).

Figure 10.1   The basic BPMN shapes.

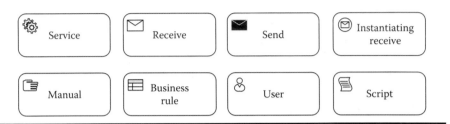

Figure 10.2   Types of activities.

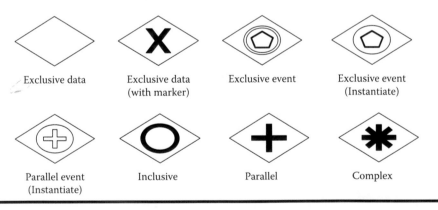

Figure 10.3   Types of gateways.

Figure 10.4   Types of events.

Figure 10.5   Types of flows.

10.2 Technique

The basic shapes, and their variations, are almost self-explanatory. It helps to have a fairly solid business process description, either written somewhere, or very well understood. The most interesting part of BPMN is the possibility of pools and swim lanes. Create a pool by putting a swim lane inside another swim lane. Presumably, a pool has more than one swim lane, and these can nicely correspond to organizational groups. Both pools and swim lanes can be named. Use a separate pool/swim lane for external groups. Activities and decisions inside a swim lane are very much like flowcharts. A "well-formed" BPMN should have a single start event and a single end event, although there can be internal activities inside swim lanes. Also, major activities can be decomposed into separate BPMN diagrams, in much the same way that flowchart processes can be decomposed.

10.3 Example

The simplest view of a BPMN model is that it is an enriched flowchart that shows the flow of information, communication paths, and data among components of an organization. As a mild departure from the previous chapters, Figure 10.6 models a typical organizational response to a field trouble report.

10.4 Deriving Test Cases from a Business Process Modeling and Notation Definition

10.4.1 The Insurance Premium Problem

Figure 10.7 is an attempt to show the Insurance Premium Problem as a business process. The two parties, the person to be insured and the insurance company, are portrayed as separate swim lanes. This doesn't change the test generation aspect of MBT, but it does indicate communication paths that would need to be checked (probably manually).

The BPMN notation allows activities to be decomposed in a separate diagram. The Compute premium activity would have a BPMN diagram very similar to that of the flowchart model in Chapter 2 (see Figure 2.9). Test case generation will be nearly identical to that for the flowchart model in Chapter 2. (See Chapters 14 and 18 for test cases generated by commercial MBT products from BPMN models.)

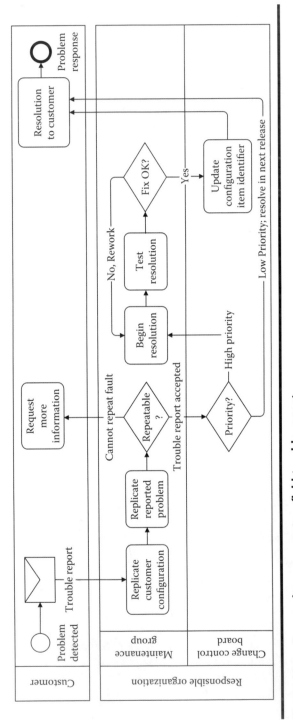

Figure 10.6 Process for response to a field trouble report.

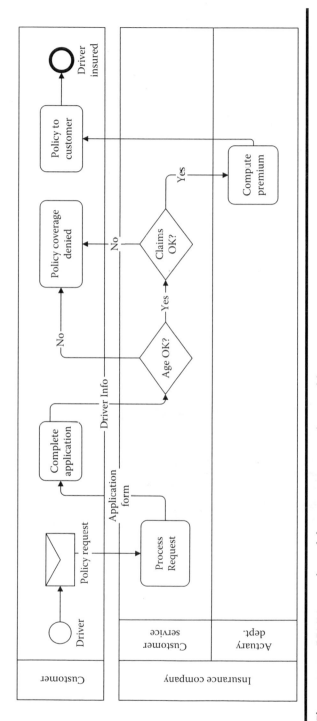

Figure 10.7 BPMN overview of the Insurance Premium Problem.

10.4.2 The Garage Door Controller

Due to the extensive similarity to both flowcharts and UML Activity Diagrams, there is no point in trying to model an event-driven system with BPMN.

10.5 Advantages and Limitations

BPMN is clearly helpful to define and describe business processes. Pools and swim lanes are very handy, particularly to describe "who does what, and when." The varieties of activities and gateways constitute a very rich symbolic "vocabulary" with which to describe business processes. The little icons in the upper left corner of the activities have some mnemonic value; however, the eight gateways are less helpful. As with other rich models, the possibility of exact expression comes at the price of mastering the vocabulary.

As a tool for model-based testing, BPMN is of limited value. Can it be done? Certainly, commercial MBT products demonstrate this. Is it a good idea? Probably not; flowcharts and decision tables are better choices.

This chapter concludes with the continuing table of behavioral issue representation (Table 10.1).

Table 10.1   Representation of Behavioral Issues with BPMN Diagrams

Issue	Represented?	Comment
Sequence	Yes	As with flowcharts
Selection	Yes	As with flowcharts
Repetition	Yes	As with flowcharts
Enable	No	Must describe as text in an activity box
Disable	No	Must describe as text in an activity box
Trigger	No	Must describe as text in an activity box
Activate	No	Must describe as text in an activity box
Suspend	No	Must describe as text in an activity box
Resume	No	Must describe as text in an activity box
Pause	No	Must describe as text in an activity box
Conflict	No	Must describe as text in an activity box
Priority	No	Must describe as text in an activity box

(Continued)

Table 10.1 (*Continued*)   Representation of Behavioral Issues with BPMN Diagrams

Issue	Represented?	Comment
Mutual exclusion	Yes	Careful use of gateway events
Concurrent execution	Partial	Must describe as text in an activity box
Deadlock	No	
Context-sensitive input events	No	
Multiple-context output events	No	
Asynchronous events	No	
Event quiescence	No	
Memory?	Yes	By reference to "previous" decisions, inputs, and activity boxes
Hierarchy?	Yes	Activity boxes and be expanded into more detail as needed

THE PRACTICE OF MODEL-BASED TESTING

2

Chapter 11

About the International Software Testing Qualification Board

According to its website (www.istqb.org) the International Software Testing Qualification Board (ISTQB®) *"is the de facto standard in certification of software testing competences."* The ISTQB offers three levels of software testing certification—Foundation, Advanced, and Expert. New modules are being added in four topic areas—agile, automotive, usability, and model-based testing (MBT). As of June 2015, the ISTQB had certified 410,803 testers in more than 100 countries worldwide.

11.1 The ISTQB Organization

The ISTQB is a nonprofit organization, it is composed of 50 member boards (these are generally country-specific) that serve 72 countries, and has accredited more than 250 training providers. The vision statement (text quoted from the ISTQB website is in italic font):

> *"To continually improve and advance the software testing profession by:*
>
> - *Defining and maintaining a Body of Knowledge which allows testers to be certified based on best practices,*
> - *connecting the international software testing community,*
> - *and encouraging research."*

The mission statement, also quoted from the ISTQB website is

1. *"We promote the value of software testing as a profession to individuals and organizations.*
2. *We help software testers to be more efficient and effective in their work, through the certification of competencies.*
3. *We enable testers to progress their career through a Professionals' Code of Ethics and a multi-level certification pathway that provides them with the skills and knowledge they need to fulfil their growing responsibilities and to achieve increased professionalism.*
4. *We continually advance the Testing Body of Knowledge by drawing on the best available industry practices and the most innovative research, and we make this knowledge freely available to all.*
5. *We set the criteria for accrediting training providers, to ensure consistent delivery of the Body of Knowledge, world-wide.*
6. *We regulate the content and coverage of exam questions, the examination process, and the issuing of certifications by official examination bodies.*
7. *We are committed to expanding software testing certifications around the world, by admitting member boards into the ISTQB®. These boards adhere to the constitution, bylaws, and processes defined by the ISTQB®, and participate in regular audits.*
8. *We nurture an open international community, committed to sharing knowledge, ideas, and innovations in software testing.*
9. *We foster relationships with academia, government, media, professional associations and other interested parties.*
10. *We provide a reference point against which the effectiveness of testing services can be evaluated, by maintaining our prominence as a respected source of knowledge in software testing."*

As an all-volunteer organization, with the exception of paid staff at the world headquarters in Belgium and in the country offices, the ISTQB is governed by an Executive Committee (president, vice president, treasurer, and secretary) elected by the General Assembly, which is composed of representatives from the member boards. The work is done by Working Groups that are chartered by the General Assembly. The MBT extension to the Foundation Level Syllabus is one such working group. Members of the working groups are generally well recognized both within their respective countries and also internationally.

11.2 Certification Levels

ISTQB certification is done at three levels: Foundation, Advanced, and Expert. At the Foundation Level there are existing extensions: Agile Tester and Model-Based Tester. In addition, there are planned extensions for Usability Tester and Automotive

Tester. The Advanced Level certification is divided into three roles—Test Manager, Test Analyst, and Technical Test Analyst. As with the Foundation Level, there are two agile extensions—Organizational and Technical. On the Specialist side, there are planned extensions for Security Tester and Test Automation Engineer. The Expert Level has two certifications—Test Management and Improving the Test Process. Syllabi for all of these certifications are available for (free) download at the ISTQB website.

11.3 The ISTQB MBT Syllabus

The model-based testing extension to the Foundation Syllabus was approved by the ISTQB General Assembly in October 2015 [ISTQB 2015]. The Working Group was composed of an overall chair, and cochairs for the Author, Exam Question, Exam Question Reviewer, and an overall Review group. There were eight authors, twelve exam authors, four exam question reviewers, and thirteen members of the review group.

Each ISTQB syllabus has stated Learning Objectives, Expected Business Outcomes, and recommended times for training providers. The recommended training times fit into 12 hours of instruction, presumably spread over a two-day course. The ISTQB-stated Learning Objectives are quoted directly from the syllabus. Each Learning Objectives is linked to the section in this book where the material is covered.

11.3.1 Introduction to Model-Based Testing

This section describes the primary reasons that any organization should consider using MBT—effectiveness and efficiency. Both are consequences of the early modeling activity that improves communication among all the stakeholders in the development process. The automation of test execution supports more thorough testing (because the tests are derived from the models) and also realizes the long-sought goal of requirements tracing. The section cautions that MBT is not a cure-all for poor past testing performance. The effort must go into the models, and generated (and automatically executed) tests are only as good as the models from which they are derived. The introduction of MBT technology inevitably forces changes in the adopting organization's development process; the first section concludes with experience-based advice. The stated Learning objectives are as follows:

"11.3.1.1 Objectives and Motivations for MBT

- *Describe expected benefits of MBT*
- *Describe misleading expectations and pitfalls of MBT*

11.3.1.2 MBT Activities and Artifacts in the Fundamental Test Process

- *Summarize the activities specific to MBT when deployed in a test process*
- *Recall the essential MBT artifacts (inputs and outputs)*

11.3.1.3 Integrating MBT into the Software Development Lifecycle

- *Explain how MBT integrates into software development lifecycle processes*
- *Explain how MBT supports requirements engineering"*

11.3.2 Model-Based Testing Modeling

It is no great surprise that the section on modeling is the most comprehensive part of the syllabus. Most of this section of the syllabus speaks "about" modeling—there is very little actual modeling instruction. (In contract, Part 1 of this book provides very comprehensive explanations of several different models, and advice on how best to use them.) Two example models are provided in Appendix A. One is of a UML Activity Diagram with generic symbols. It does not represent a particular application. The second example is presented as a state machine, but it violates the usual finite state machine formalisms by adding decision boxes. It is more like a blend of UML Activity Diagrams and traditional finite state machines. The section concludes with helpful advice about requirements tracing and the influence of MBT tools.

"11.3.2.1 Model-Based Testing Modeling

- *Develop a simple MBT model for a test object and predefined test objectives using a workflow-based modeling language (refer to Section 8.1—"simple" means less than 15 modeling elements)*
- *Develop a simple MBT model for a test object and predefined test objectives using a state transition-based modeling language (refer to Section 8.2—"simple" means less than 15 modeling elements)*
- *Classify an MBT model with respect to the subject and to the focus*
- *Give examples of how an MBT model depends on the test objectives*

11.3.2.2 Languages for MBT Models

- *Recall examples of modeling language categories commonly used for MBT*
- *Recall typical representatives of modeling language categories relevant for different systems and project objectives*

11.3.2.3 Good Practices for Model-Based Testing Modeling Activities

- *Recall quality characteristics for MBT models*
- *Describe classic mistakes and pitfalls during modeling activities for MBT*
- *Explain the advantages of linking requirements and process related information to the MBT model*
- *Explain the necessity of guidelines for MBT modeling*
- *Provide examples where reuse of existing models (from requirements phase or development phase) is or is not appropriate*
- *Recall tool types supporting specific MBT modeling activities*
- *Summarize iterative MBT model development, review, and validation"*

11.3.3 Selection Criteria for Test Case Generation

MBT tools, particularly the commercial ones, can generate large numbers of test cases. In an automated test environment, this is not a problem. The number of generated test cases can usually be reduced with the careful use of selection criteria, for example, coverage of some aspect of the underlying model. These criteria can often be supplemented by organizational priorities, such as project risk, or strong safety requirements. Many of these capabilities are MBT product-specific, and these are important in eventual tool selection.

"11.3.3.1 Classification of Model-Based Testing Test Selection Criteria

- *Classify the various families of test selection criteria used for test generation from models*
- *Generate test cases from an MBT model to achieve given test objectives in a given context*
- *Provide examples of model coverage, data-related, pattern- and scenario-based and project-based test selection criteria*
- *Recognize how MBT test selection criteria relate to ISTQB Foundation Level test design techniques*

11.3.3.2 Applying Test Selection Criteria

- *Recall degrees of test artifact generation automation*
- *Apply given test selection criteria to a given MBT model*
- *Describe good practices of MBT test selection criteria"*

11.3.4 MBT Test Implementation and Execution

This is the second largest section of the syllabus. It clarifies distinctions between abstract and concrete test cases, and describes three levels of MBT test execution: strictly manual, strictly automated, and some combinations of these. Automated test execution requires some form of an "oracle" that describes expected outcomes, and is the basis for pass/fail decisions made by the execution tool. Much of the utility of an oracle depends on the underlying model. For example, if the SUT is modeled as an finite state machine, the events and conditions that cause a transition are indicated as well as expected outcome actions.

"11.3.4.1 Specifics of Model-Based Testing Test Implementation and Execution

- *Explain the difference between abstract and concrete test cases in the MBT context*
- *Explain the different kinds of test execution in the MBT context*
- *Perform updates of an MBT model and test generation caused by changes in requirements, test objects, or test objectives*

11.3.4.2 Activities of Test Adaptation in Model-Based Testing

- *Explain which kind of test adaptation may be necessary for test execution in MBT"*

11.3.5 Evaluating and Deploying a Model-Based Testing Approach

This section covers the usual considerations that drive technology adoption:

- A perception of need (or existing weakness)
- The need to become acquainted with MBT tool capabilities
- A trial use
- A transition strategy

This whole topic is the subject of Chapter 12 of this book.

In my personal testing experience (telephone switching systems in the early 1980s), our laboratory found that even manually executed tests that were derived from models greatly improved the testing process and overall end quality. When we built an automated regression testing system (I was part of the team), the newly found freedom resulted in testers creating many more tests than they would have in a manual execution situation.

Most of the ISTQB Learning Objectives are covered elsewhere in this book, as shown in Table 11.1.

Table 11.1   Mapping ISTQB Learning Objectives to This Book

ISTQB Objectives Category	Specific Objectives	Section/ Chapter
1.1 Objectives and motivations for MBT	Describe expected benefits of MBT	Section 1.9, Chapters 14 through 19
	Describe misleading expectations and pitfalls of MBT	Section 1.9, Chapters 14 through 19
1.2 MBT activities and artifacts in the fundamental test process	Summarize the activities specific to MBT when deployed in a test process	Section 1.9
	Recall the essential MBT artifacts (inputs and outputs)	Section 1.9
1.3 Integrating MBT into the software development lifecycle	Explain how MBT integrates into software development lifecycle processes	Chapters 2 through 10
	Explain how MBT supports requirements engineering	Chapters 2 through 10
2.1 MBT modeling	Develop a simple MBT model for a test object and predefined test objectives using a workflow-based modeling language (refer to Section 8.1—"simple" means less than 15 modeling elements)	Chapters 2 through 10
	Develop a simple MBT model for a test object and predefined test objectives using a state transition-based	Chapters 2 through 10
	Classify an MBT model with respect to the subject and to the focus	Chapters 2 through 10
	Give examples of how an MBT model depends on the test objectives	Chapters 2 through 10

(Continued)

Table 11.1 (*Continued*)   Mapping ISTQB Learning Objectives to This Book

ISTQB Objectives Category	Specific Objectives	Section/ Chapter
2.2 Languages for MBT models	Recall examples of modeling language categories commonly used for MBT	Chapters 2 through 10
	Recall typical representatives of modeling language categories relevant for different systems and project objectives	Chapters 2 through 10
2.3 Good practices for MBT modeling activities	Recall quality characteristics for MBT models	Chapters 2 through 10
	Describe classic mistakes and pitfalls during modeling activities for MBT	Chapters 2 through 10
	Explain the advantages of linking requirements and process related information to the MBT model	Chapter 1
	Explain the necessity of guidelines for MBT modeling	—
	Provide examples where reuse of existing models (from requirements phase or development phase) is or is not appropriate	—
	Recall tool types supporting specific MBT modeling activities	Chapters 2 through 10
	Summarize iterative MBT model development, review, and validation	—
3.1 Classification of MBT test selection criteria	Classify the various families of test selection criteria used for test generation from models	Chapters 2 through 10
	Generate test cases from an MBT model to achieve given test objectives in a given context	Chapters 2 through 10
	Provide examples of model coverage, data-related, pattern- and scenario-based and project-based test selection criteria	Chapters 14 through 19
	Recognize how MBT test selection criteria relate to ISTQB Foundation Level test design techniques	—

(Continued)

Table 11.1 (*Continued*) Mapping ISTQB Learning Objectives to This Book

ISTQB Objectives Category	Specific Objectives	Section/ Chapter
3.2 Applying test selection criteria	Recall degrees of test artifact generation automation	Chapter 1
	Apply given test selection criteria to a given MBT model	Chapters 14 through 19
	Describe good practices of MBT test selection criteria	Chapters 14 through 19
4.1 Specifics of MBT test implementation and execution	Explain the difference between abstract and concrete test cases in the MBT context	Chapter 1
	Explain the different kinds of test execution in the MBT context	Chapter 10
	Perform updates of an MBT model and test generation caused by changes in requirements, test objects, or test objectives	—
4.2 Activities of test adaptation in MBT	Explain which kind of test adaptation may be necessary for test execution in MBT	—
5.1 Evaluate an MBT deployment	Describe ROI (Return On Investment) factors for MBT introduction	Chapter 1
	Explain how the objectives of the project are related to the characteristics of the MBT approach	—
	Recall selected metrics and key performance indicators to measure the progress and results of MBT activities	—
5.2 Manage and monitor the deployment of an MBT approach	Recall good practices for test management, change management, and collaborative work when deploying MBT	—
	Recall cost factors of MBT	—
	Give examples of the integration of the MBT tool with configuration management, requirements management, test management, and test automation tools	—

Reference

[ISTQB 2015]
International Software Testing Certification Board (ISTQB), *Certified Tester Foundation Level Specialist Syllabus Model-Based Tester*, Available at istqb.org, October 2015.

Chapter 12

Implementing MBT in an Organization

Model-based testing (MBT) is a major change, and probably qualifies as a true paradigm shift, at least if the implementation in an organization is done correctly. This chapter is an echo of decades of research and experiences with technology adoption. For an excellent and seminal read on this, see [Bouldin 1989]. Barbara Bouldin was an employee of AT&T Bell Labs, Murray Hill, New Jersey and she was writing about the process of introducing Computer-Aided Software Engineering (CASE) technology into a large software development organization. To get a flavor of her book, these are her chapter titles (some shortened):

1. Assessing the need
2. Selecting candidate products
3. Evaluating the candidate products
4. Presenting the product to upper management
5. Presenting the product to users
6. Gathering information
7. The pilot team
8. Planning the introduction
9. Standards and naming conventions
10. Implementing change
11. Finishing the implementation
12. Finishing the implementation—averting disasters
13. Finishing the implementation—measuring the benefit
14. Finishing the implementation—handling success

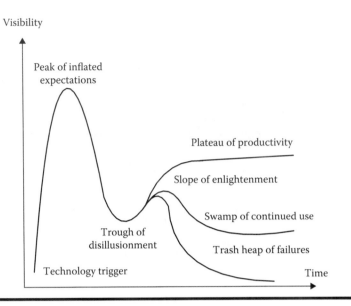

Figure 12.1   Gartner Group hype cycle.

Figure 2.11 in Chapter 2 shows the Gartner Group summary of technology adoption. It is repeated here as Figure 12.1 for convenience. The material in this chapter explains why these peaks and troughs occur, and some ideas on how to either manage or avoid them.

12.1 Getting Started

In many ways, organizational change is similar to human change. We know about 12-step programs for substance abuse—maybe this chapter should be a 12-step program for MBT adoption. Some of the substance abuse steps are clearly related to a higher power, and do not apply here. The three steps that pertain to organizational change are repeated below. The quoted part in italic font is from [Alcohol 2016]; each of these steps is rewritten to refer to MBT adoption.

> *"1. We admitted we were powerless over alcohol—that our lives had become unmanageable."*

We recognize that our testing process is inadequate; it is difficult to manage, and can be chaotic. This affects our products, our development process, our reputation, and our staff satisfaction.

> *"4. Made a searching and fearless moral inventory of ourselves."*

We will examine our testing process, tools, and staff training with honesty and no recrimination.

> *"10. Continued to take personal inventory, and when we were wrong, promptly admitted it."*

We will continue to examine our testing process, tools, and staff training during and after the transition to model-based testing.

12.1.1 Recognizing the Need for Change

Any attempt to change, whether for substance abuse, or for process improvement, must begin with a recognition to the need for a change. Many software developments end up squeezing testing into a shortened interval due to pressing deadlines. This is one indicator of a need for change. Another likely indicator is the frequency, and severity, of field trouble reports. The MBT surveys (in Chapter 1) provide a variety of reasons about why an organization might want to adopt MBT. Here are some others:

- Seems like a good idea
- Progressive companies are trying it
- Measuring our testing process shows the need to improvement
- Continuing problems with software quality

Process change is easier in an organization with a defined software process, probably at CMMI level 3 or higher. Such an organization has measured its testing process, and this recognition is strong motivation for change. Organizations at CMMI levels 1 and 2 can certainly be aware of their test process deficiencies, but they probably are not quantized.

The "iron triangle" of project management constraints is shown in Figure 12.2. Among the three choices, better, faster, cheaper, once an organization chooses any two, the third is determined. These constraints also apply to process change—in our case MBT adoption (and transition). In the first MBT survey (see Chapter 1) [Binder 2012], respondents reported that all three can be obtained; there was no information about the other pair of constraints:

> *Better: "On average, respondents report MBT reduced escaped bugs by 59%"*
> *Faster: "On average, respondents report MBT reduced testing duration by 25%"*
> *Cheaper: "On average, respondents report MBT reduced testing costs by 17%"*

The more important point is that all three vertices of the iron triangle are attainable with MBT. This is supported by the 2014 MBT survey (also in Chapter 1) [Binder 2014]. Among 100 respondents, the reported benefits of MBT included: better test

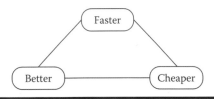

Figure 12.2 The *iron triangle*.

coverage, more control of complexity (the Better vertex), automatic test case generation, and reuse of models and model elements (the Faster vertex). Most significant
reported obstacles (read not cheaper) were tool support, skills availability for MBT,
and resistance to change. The iron triangle strikes again!

Among the respondents, the expectations included more efficient test design,
more effective test cases, managed complexity of system testing, improved communication, and the possibility to start test design earlier. Quoting from the report,
these expectations were as follows:

"Overall effectiveness of MBT

- *23.6% extremely effective*
- *40.3% moderately effective*
- *23.6% slightly effective*
- *5.6% no effect*
- *1.4% slightly ineffective*
- *2.8% moderately ineffective*
- *2.8% extremely ineffective"*

The conclusion to all of this is that, when an organization considers introducing
MBT, it should first decide its overall goal: better (overall improved testing), faster
(automated test case generation and execution), or cheaper (not likely at first). As
with most other technology changes (see Figure 12.1), in the long run, the new
technology results in lower overall cost.

12.1.2 Technology Champions

In the process of technology change, usually one person is particularly passionate
about the new technology. This person/role has become known as the "Technology
Champion." Technology champions generally come from the technical side of an
organization, but there is no reason why they could not come from the management side. Larger organizations may have a process improvement group, and that
would be a likely home for a technology champion.

Regardless of the organizational origin, the technology champion has a central
role in the technology transition. In terms of Barbara Bouldin's chapter titles,
the technology champion would almost certainly be involved in the first two

steps—need assessment and identifying relevant MBT products. The "technology trigger" in Figure 12.1 is where the MBT champion would start.

The first task of the MBT champion is to capture the interest, and approval, of both the management side and the technical side of the organization. Often referred to as "buy in," this is an early commitment by both management and development to investigate the new technology—in this case, MBT.

12.2 Getting Started

Beginning anything is usually the hard part. (I postponed writing Chapter 1 until after most of the remaining chapters were done.) The same is true for MBT adoption. Presumably, the MBT champion knows why the organization is even considering MBT—better testing, more efficient testing, or less expensive testing. In a perfect organization, there would be measurements of the testing process. At this point, the need is known, and there is initial "buy-in" to the idea. The organization is on the steeply rising part of the technology curve in Figure 12.1—the expectations.

12.2.1 Candidate Model-Based Testing Products

The first choice the MBT champion must make is between commercial products and the lower-cost (often free) open-source MBT products. There is usually very little or even no support for and from the open-source products. Many times, they are discontinued without notice—this happened to the Microsoft Spec Explorer product. Chapters 14 through 19 were provided by six MBT vendors; they are included here to help in the selection of candidate MBT products. The commercial vendors are very helpful (for obvious reasons) to organizations considering MBT adoption. Many offer consulting and coaching support services. Given the choice between open-source and commercial products, I strongly favor the commercial option.

The MBT champion clearly knows the kind of software on which MBT will be used. David Harel refers to two fundamental types—transformational and reactive [Harel 1988]. Transformational programs, as the name implies, simply transform their inputs into outputs. COBOL programs are a good, but not the only, example of transformational programs. Usually, they run and then finish. Reactive programs, in Harel's words, are longer-running programs, often event-driven, that maintain a prolonged connection to their environment, reacting to input events as they occur. Industrial control systems and avionic systems are good examples of reactive programs. The Insurance Premium Program is transformational; the Garage Door Controller is reactive.

12.2.2 Success Criteria

Introduction of any new technology is an expensive process. The MBT champion should identify agreed-upon success criteria early in the MBT investigation.

This will help reduce the height of the "peak of inflated expectations" in Figure 12.1. Section 11.3.5 of the ISTQB MBT syllabus presents generic success criteria and a summary of relevant considerations [ISTQB 2015]. An organization should consider both initial and continuing costs, including initial planning, exploration costs, product choice, process definition and change, and staff training.

Generic success criteria are just a starting point—an organization needs to define its own criteria based on company-specific priorities, deficiencies, and other local considerations. This is another point at which the better–faster–cheaper triangle is relevant.

12.2.3 Pilot Projects

The pilot MBT project is a critical step—if the example is too difficult, the result can be an undeserved failure; if it is too simple, it will not be persuasive to both the management and the technical sides of the organization. Pilot projects should be typical of the kind of software the organization builds and tests. If there are several distinct types, it may be helpful to have more than one pilot project. Generally speaking, a pilot project should be typical, but not too large. The sepp.med GmbH (see Chapter 18) has an interesting approach to pilot projects. They prefer a pilot project with medium complexity and assign it to two teams that work in parallel. One team follows the existing test process, identifying test cases manually. The other team uses the MBT approach, and is helped by vendor consultants. The activities of each team are carefully tracked, resulting in an interesting and direct comparison between the old and new ways with respect to effort, effectiveness of the test cases, and the overall quality of the generated tests. This is a little artificial due to the vendor participation; on the other hand, it is a good indicator of eventual success (the "plateau of productivity" in Figure 12.1).

12.3 Training and Education

Many people, particularly university types, try to distinguish the difference between training and education. Training is usually seen as product-specific, whereas education is usually seen as more durable, more general. My favorite illustration asks whether sex training or sex education should be taught in public schools. Now we all know the difference! Chapters 2 through 10 are intended to be educational; to some extent, Chapters 14 through 19 are training oriented. All the commercial MBT vendors offer customer training, and most offer free short-term product licenses for pilot projects. Every MBT vendor will assert that the success of MBT is only as good as the models used by the product. The hardest part of a pilot project is learning how to use candidate product well. This is where the sep.med GmbH idea is appropriate. If an organization is evaluating more than one MBT product, the learning curves quickly add up. That is part of the hope for Chapters 14 through

19—they give an idea of what the individual products do. The reason I asked vendors to provide individual chapters is that I did not want to misrepresent product capabilities. Also (true confession) I did not have to go through the learning curve on six products.

The main cause for the "trough of disillusionment" in Figure 12.1 is the lack of education in modeling techniques, and a corresponding lack of training in an eventually selected MBT product. The "Slope of Enlightenment" part of the curve is due to both training and education. The "trash heap of failures" occurs for two reasons: maybe the technical staff does not understand the modeling techniques, or maybe they do not know how to use the MBT product well (or both). The "swamp of continued use" occurs when there is just enough training and education to use the product, but not enough of either one to use it well. The title of this book reflects this—in order for model-based testing to be a craft, the modeling technology and the product technology must be understood. Continued education and training will sustain the slope of the "Slope of Enlightenment" until eventually, there is a point of stability.

12.4 Lessons Learned from Experience

In my 20-year career in the telephone switching systems business, I witnessed many introductions of new technologies. I described one of them in Chapter 2 with the AELFlow automated flowchart generator. Barbara Bouldin's book is right on target, particularly the part about administrative and technical "buy-in". In the R&D laboratory where I worked, there was an overly enthusiastic tools group that wanted to develop tools for everything they could think of. The organization developed a policy that, before a new tool could be developed internally, the underlying technique/technology had to be used on a pilot project. Many times, this acted as a "rapid prototype" and actually helped the proposed tool. The value of a well-chosen pilot project cannot be minimized. It must be sufficiently complex to persuade any doubter in the technical community, yet it cannot be too large (expensive) to minimize risk.

My second major experience with technology introduction was with the CASE "revolution" in the early 1990s. When I came to my present university, I had a half-time appointment in the newly-opened Research and Technology Institute of West Michigan (RTI). My job description was simple—put Grand Rapids on the "technical community map." Based on my industrial experience, I established the RTI CASE Tool Laboratory that, at its peak, contained 26 prominent CASE tools. The premise of the lab was simple—reduce/eliminate the learning curve on CASE tools to accelerate the tool adoption process. Clients could use the lab, with our staff that knew the tools, and try out a variety of CASE tools without going through the learning process on each tool. This part worked well, but when I recall failed CASE transitions, without exception, the failures were due to the lack of training in the techniques supported by the tools. One of the early CASE commentators was often quoted, "A fool with a tool is still a fool." This is a little harsh, but the underlying point also applies to MBT adoption. The whole point of Chapters 2 through 10 is to provide at least a start in knowing how to do good modeling.

The title of this book, *The Craft of Model-Based Testing*, was chosen for the continuing analogy between modeling as a craft and more traditional forms of craft. When I have time, I enjoy woodworking. Becoming proficient in any craft, three things must be understood—the medium, appropriate tools, and being able to use the tools well.

12.4.1 The Medium

In the case of woodworking, the would-be craftsperson needs to know the characteristics of the medium—the wood. Hard versus soft wood, close grained versus wide grained, and so on. Oak is quite hard, and very strong, but harder to work with. Maple is even worse. My favorite is cherry wood—it is not as hard as oak or maple, yet it can be worked well, and takes a fine finish.

In the MBT domain, the would-be craftsperson needs to understand the nature of the software being tested. Is it transformational, or reactive, data-driven or event-driven, and so on. These differences imply good and poor choices of both tools and techniques. In 1981, James Peterson described a hierarchy of system complexity [Peterson 1981], shown in Figure 12.3. Figure 12.4 is an update showing work completed after Peterson's book.

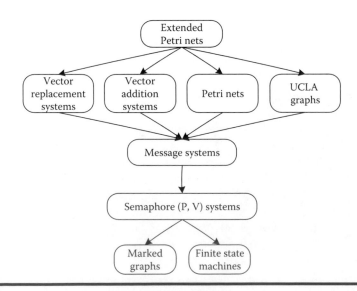

Figure 12.3   Peterson's lattice of control models.

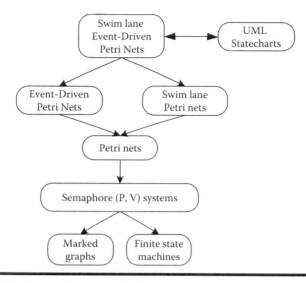

Figure 12.4 Update to Peterson's lattice.

Three of the deleted models, Vector Replacement Systems, Vector Addition Systems, and UCLA graphs, are no longer active, even in academic circles. Semaphore systems are still supported by a few programming languages, and marked graphs are most frequently seen as the Yourdon-style dataflow diagrams. Most commercial tools are based on finite state machines, although there are a few that claim to support Statecharts.

Each of the modeling chapters in Part 1 ends with a table showing the expressive capabilities of the model described in that chapter. The best way to understand the medium is to consider which of the 19 behavioral considerations pertain to the problem at hand, and use the table to determine whether that model has sufficient expressive capability. If a problem demands a model of the complexity of Petri nets (or more), a tool that only supports finite state machines, will not be able to express the underlying application completely.

12.4.2 *The Tools*

As a woodworker, I have more tools that I regularly use. For example, I have seven special-purpose hand saws, and as a would-be craftsman, I at least know which tool will work best in a given situation. There is an axiom among woodworkers—a craftsman can do reasonable work with inferior tools, but good tools will not turn a neophyte into a craftsperson.

Vendors are usually very generous with temporary licenses of their products, but taking advantage of this will always require vendor-supplied tool training. Chapters 14 through 19 present vendor-provided information on six commercial

MBT products. This is clearly a good place to become familiar with capabilities of available MBT products.

12.4.3 The Ability to Use Tools Well

As a woodworker, I sometimes try a new tool on some scrap wood before using it on a real project. Using any tool well requires practice. The traditional paradigm of Master, Journeyman, and Apprentice was probably the epitome of the need for supervised practice.

Using an MBT tool well requires both training in the tool and education in the underlying modeling technique. Many MBT vendors offer both training and coaching on real projects. One company estimates that in 80 hours, a designer can become MBT proficient. To have a successful transition to MBT, an organization must plan for both tool training and modeling education. The specter of CASE failures almost guarantees this.

References

[Alcohol 2016]
Alcoholics Anonymous World Services, *Alcoholics Anonymous*, 4th ed., ISBN 1-893007-16-2, in PDF format, (free download) 2016.
[Binder 2012]
Binder, Robert V., Real Users of Model-Based Testing, blog, http://robertvbinder.com/real-users-of-model-based-testing/, January 16, 2012.
[Binder 2014]
Binder, Robert V., Anne Kramer, and Bruno Legeard, *2014 Model-based Testing User Survey: Results*, 2014.
[Bouldin 1989]
Bouldin, Barbara M., *Agents of Change: Managing the Introduction of Automated Tools*. Yourdon Press, Upper Saddle River, NJ, 1989.
[Harel 1988]
Harel, David, On visual formalisms. *Communications of the ACM* 1988;31(5):514–530.
[ISTQB 2015]
International Software Testing Certification Board (ISTQB), *Certified Tester Foundation Level Specialist Syllabus Model-Based Tester*, available at istqb.org, October 2015.
[Peterson1981]
Peterson, James L., *Petri Net Theory and the Modeling of Systems*. Prentice Hall, Englewood Cliffs, NJ, 1981.

Chapter 13

Information Provided to Model-Based Testing Tool Vendors

The purpose of the chapters in Part 2 is to provide a side-by-side comparison of commercial and open-source model-based testing (MBT) tools. The vendors of commercial tools were given the two continuing problem statements (the Insurance Premium Problem and the Garage Door Controller), along with selected models from the relevant chapters in Part 1. The comments on the open-source MBT tools in Chapter 20 are from graduate student projects at Grand Valley State University, Allendale, Michigan. Chapters 14 through 19 follow the template below.

13.1 Chapter Template

The vendor-provided materials will be placed into the template below to the extent possible. This is intended to assist in any vendor comparisons.

Chapter xx   <Vendor Name>
xx.1   Introduction
xx.1.1   About the Vendor
xx.1.2   About the Vendor Product(s)
xx.1.3   Customer Support

xx.2   Insurance Premium Results
xx.2.1   Problem Input

xx.2.2   Generated Test Cases
xx.2.2.1   Abstract test cases generated by <tool name, model used>
xx.2.2.2   Concrete test cases generated by <tool name, model used>
xx.2.3   Other Vendor-Provided Analysis

xx.3   Garage Door Controller Results
xx.3.1   Problem Input
xx.3.2   Generated Test Cases
xx.3.2.1   Abstract test cases generated by <tool name, model used>
xx.3.2.2   Concrete test cases generated by <tool name, model used>
xx.3.3   Other Vendor-Provided Analysis

xx.4   Vendor Advice

In this section, MBT tool vendors are invited to make suggestions about training that may be needed, introduction/transition strategies, client experience(s), and similar information. No pricing information is permitted—the purpose is for comparison and not for advertising.

13.2 A Unit Level Problem: Insurance Premium Calculation

13.2.1 Problem Definition

Premiums on an automobile insurance policy are computed by cost considerations that are applied to a base rate. The inputs to the calculation are

1. The base rate ($600)
2. The age of the policy holder ($16 <=$ age $< 25; 25 <=$ age $< 65; 65 <=$ age < 90)
3. People less than age 16 or more than 90 cannot be insured
4. The number of "at fault" claims in the past 5 years (0, 1 to 3, 3 to 10)
5. Drivers with more than 10 at fault claims in the past 5 years cannot be insured
6. The reduction for being a goodStudent ($50)
7. The reduction for being a nonDrinker ($75)

The calculation values are shown in Tables 13.1 through 13.3.

13.2.2 Problem Models

13.2.2.1 Flowchart

Figure 2.7 is a flowchart model of the Insurance Premium Calculation as defined in Chapter 1, Section 1.8.1 and repeated here as Figure 13.1.

Table 13.1   Premium Multiplication Values for Age Ranges

Age Ranges	ageMultiplier
16 <= age < 25	x = 1.5
25 <= age < 65	x = 1.0
65 <= age < 90	x = 1.2

Table 13.2   Premium Penalty Values at Fault Claims

"At Fault" Claims in Past 5 years	claimsPenalty
0	$0
1 to 3	$100
4 to 10	$300

Table 13.3   Decision Table for goodStudent and nonDrinker Reductions

c1. goodStudent	T	T	F	F
c2. nonDrinker	T	F	T	F
a1. Apply $50 reduction	x	x	—	—
a2. Apply $75 reduction	x	—	x	—
a3. Do nothing	—	—	—	—

13.2.2.2 Decision Table(s)

The definition of the Insurance Premium Problem (Chapter 1) supports nearly direct development of a Mixed Entry Decision Table (MEDT). The age and "at fault claims" variables have defined ranges that lead directly to the extended entry conditions c1 and c2 in Table 13.4. The premium reductions are both boolean, so conditions c3 and c4 are limited entry conditions. For space reasons, Table 13.4 is split into three parts. There are no impossible rules, nor are there any Do Nothing actions.

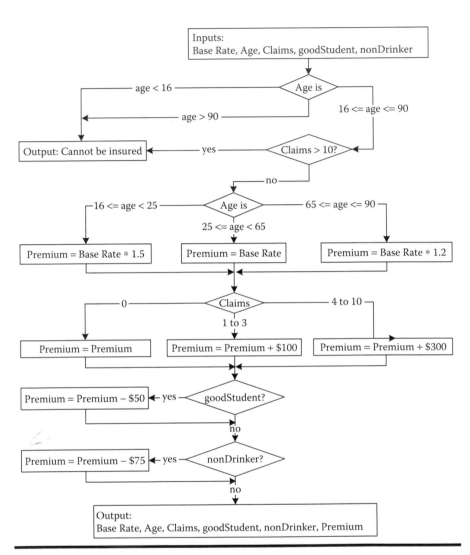

Figure 13.1   Insurance Premium Flowchart.

Table 13.4 Insurance Premium Problem Decision Table (Part 1)

	1	2	3	4	5	6	7	8	9	10	11	12
c1. Age						16 <= age < 25						
c2. At fault claims	0				1 to 3				4 to 10			
c3. Good student?	T	T	F	F	T	T	F	F	T	T	F	F
c4. Non-drinker?	T	F	T	F	T	F	T	F	T	F	T	F
a1. Base times 1.5	X	X	X	X	X	X	X	X	X	X	X	X
a2. Base times 1												
a3. Base times 1.2												
a4. add $0 to Base	X	X	X	X								
a5. add $100 to Base					X	X	X	X				
a6. add $300 to Base									X	X	X	X
a7. subtract $50 from Base	X	X			X	X			X	X		
a8. subtract $75 from Base	X		X		X		X		X		X	
Rule	1	2	3	4	5	6	7	8	9	10	11	12

Table 13.4 Insurance Premium Problem Decision Table (Part 2)

	13	14	15	16	17	18	19	20	21	22	23	24
c1. Age	25 <= age < 65											
c2. At fault claims	0				1 to 3				4 to 10			
c3. Good student?	T	T	F	F	T	T	F	F	T	T	F	F
c4. Non-drinker?	T	F	T	F	T	F	T	F	T	F	T	F
a1. Base times 1.5												
a2. Base times 1	x	x	x	x	x	x	x	x	x	x	x	x
a3. Base times 1.2												
a4. add $0 to Base	x	x	x	x								
a5. add $100 to Base					x	x	x	x				
a6. add $300 to Base									x	x	x	x
a7. subtract $50 from Base	x	x			x	x			x	x		
a8. subtract $75 from Base	x		x		x		x		x		x	
Rule	13	14	15	16	17	18	19	20	21	22	23	24

Table 13.4 Insurance Premium Problem Decision Table (Part 3)

c1. Age	65 <= age <= 90											
c2. At fault claims	0				1 to 3				4 to 10			
c3. Good student?	T	T	F	F	T	T	F	F	T	T	F	F
c4. Non-drinker?	T	F	T	F	T	F	T	F	T	F	T	F
a1. Base times 1.5												
a2. Base times 1												
a3. Base times 1.2	x	x	x	x	x	x	x	x	x	x	x	x
a4. add $0 to Base	x	x	x	x								
a5. add $100 to Base					x	x	x	x				
a6. add $300 to Base									x	x	x	x
a7. subtract $50 from Base	x	x			x				x	x		
a8. subtract $75 from Base	x		x		x		x		x		x	
Rule	25	26	27	28	29	30	31	32	33	34	35	36

Table 13.4 Insurance Premium Problem Decision Table (Part 4)

c1. Age	<16	>90	–
c2. At fault claims	–	–	>10
c3. Good student?	–	–	–
c4. Non-drinker?	–	–	–
a9. cannot be insured	x	x	X
Rule	37	38	39

13.2.2.3 Finite State Machine

Inputs	Outputs (Actions)	States
e1: Base rate	a1: Base rate × 1.5	s1: Idle
e2: 16 <= age < 25	a2: Base rate × 1.0	s2: Apply age multiplier
e3: 25 <= age < 65	a3: Base rate × 1.2	s3: Apply claims penalty
e4: Age >= 65	a4: Add $0	s4: Apply goodStudent reduction
e5: At fault claims = 0	a5: Add $100	s5: Apply nonDrinker reduction
e6: 1 <= at fault claims <= 3	a6: Add $300	s6: Done
e7: 4 <= at fault claims <= 10	a7: Subtract $0	
e8: goodStudent = T	a8: Subtract $50	
e9: goodStudent = F	a9: Subtract $75	
e10: nonDrinker = T		
e11: nonDrinker = F		

As a finite state machine, the one in Figure 13.2 is almost useless for deriving test cases. It is technically correct, but the flowchart version is more helpful. There are 36 distinct paths through this finite state machine, and they correspond directly to the 36 paths in the flowchart (Figure 13.1) and to the 36 rules in the Mixed Entry Decision Table formulation in 13..2.2.2. It as the FSM adds nothing to the simpler models.

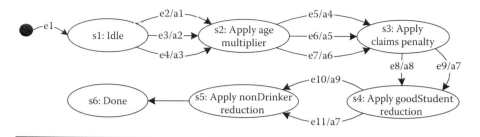

Figure 13.2 The Insurance Premium Problem finite state machine.

13.2.3 *Insurance Premium Problem Visual Basic for Applications Code*

Insurance Premium Program (Visual Basic for Applications, VBA)

```
'
Const basePrice As Currency = 600
Dim age As Integer, atFaultClaims As Integer
Dim goodStudent As Boolean, nonDrinker As Boolean
Dim premium As Currency
Input(age, atFaultClaims, goodStudent, nonDrinker)
'
'        Data validity check (Input and Output statements are not in VBA syntax)
'
If ((Age < 16) OR (age > 90))
        Then Output('Not Insured')
EndIf
If ((atFaultClaims< 0) OR (atFaultClaims> 10))
        Then Output('Not Insured')
EndIf
'        Premium calculation
premium = basePrice
Select Case age
Case 16 <= age < 25
        premium = premium * 1.5
Case 25 <= age < 65
        premium = premium
Case 65 <= age <= 90
        premium = premium * 1.2
End Select
Select Case atFaultClaims
Case 0
        premium = premium
```

Case 1 <= atFaultClaims <=3
      premium = premium + 100
Case 4 <= atFaultClaims <=10
      premium = premium + 300
End Select

If goodStudent Then premium = premium - 50
EndIf

If nonDrinker Then premium = premium - 75
EndIf

Output (age, atFaultClaims, goodStudent, nonDrinker, Premium)
End

13.3 A System Level Problem: The Garage Door Controller

13.3.1 Problem Definition

A system to open a garage door is composed of several components: a drive motor, the garage door wheel tracks with sensors at the open and closed positions, and a control device. In addition, there are two safety features: a laser beam near the floor and an obstacle sensor. These latter two devices operate only when the garage door is closing. While the door is closing, if either the light beam is interrupted (possibly by a pet) or if the door encounters an obstacle, the door immediately stops and then reverses direction. When the door is in motion, either closing or opening, and a signal from the control device occurs, the door stops. A subsequent control signal starts the door in the same direction as and when it was stopped. Finally, there are sensors that detect when the door has moved to one of the extreme positions, either fully open or fully closed. When either of these occurs, the door stops. Figure 13.3 is a SysML context diagram of the Garage Door Controller.

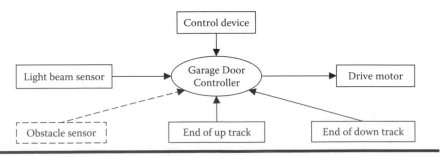

Figure 13.3 SysML diagram of the Garage Door Controller.

In most garage door systems, there are several control devices: a digital keyboard mounted outside the door, a separately powered button inside the garage, and possibly several in-car signaling devices. For simplicity, we collapse these redundant signal sources into one device. Similarly, as the two safety devices generate the same response, we will drop consideration of the obstacle sensor and just consider the light beam device.

13.3.2 Problem Models

13.3.2.1 Flowchart

Figure 13.4 is a copy of the flowchart model of the problem in Chapter 2.

13.3.2.2 Decision Table(s)

The three parts of Table 13.5 are copies of Table 3.16 in Chapter 3. The F! notation should be interpreted as "Must be False."

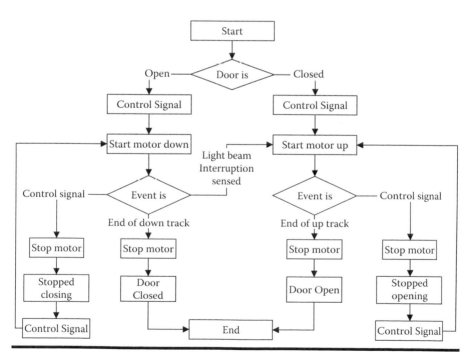

Figure 13.4 Garage Door Controller flowchart.

Table 13.5 Garage Door Controller Decision Table (Part 1)

Rule	1	2	3	4	5	6	7	8	9	10	11	12
c1. Door is	Up				Closing				Stopped going down			
c2. Control device signal	T	F!	F!	F!	T	F!	F!	F!	T	F!	F!	F!
c3. End of down track reached	F!	T	F!	F!	F!	T	F!	F!	F!	T	F!	F!
c4. End of up track reached	F!	F!	T	F!	F!	F!	T	F!	F!	F!	T	F!
c5. Light beam interruption	F!	F!	F!	T	F!	F!	F!	T	F!	F!	F!	T
a1. Start drive motor down	x								x			
a2. Start drive motor up												
a3. Stop drive motor					x	x						
a4. Reverse motor down to up				x				x				
a5. Do nothing		x	x									x
a6. Impossible							x			x	x	
a7. Repeat table	x			x	x	x		x	x			x

Note: The F! notation means *must be false.*

Table 13.5 Garage Door Controller Decision Table (Part 2)

Rule	13	14	15	16	17	18	19	20	21	22	23	24
c1. Door is	Down				Opening				Stopped going up			
c2. Control device signal	T	F!	F!	F!	T	F!	F!	F!	T	F!	F!	F!
c3. End of down track reached	F!	T	F!	F!	F!	T	F!	F!	F!	T	F!	F!
c4. End of up track reached	F!	F!	T	F!	F!	F!	T	F!	F!	F!	T	F!
c5. Light beam interruption	F!	F!	F!	T	F!	F!	F!	T	F!	F!	F!	T
a1. Start drive motor down	x											
a2. Start drive motor up									x			
a3. Stop drive motor					x		x					
a4. Reverse motor down to up												
a5. Do nothing				x		x		x				x
a6. Impossible		x	x							x	x	
a7. Repeat table	x				x				x			

Note: The F! notation means *must be false.*

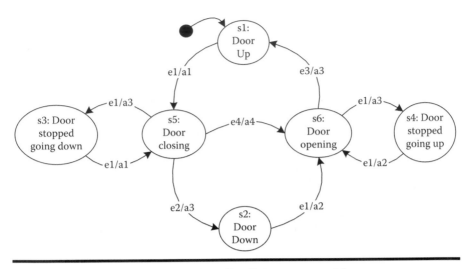

Figure 13.5 The Garage Door Controller finite state machine.

13.3.2.3 Finite State Machine

The finite state machine in Figure 13.5 is copied from Figure 4.13 in Chapter 4.

Input Events	Output Events (Actions)	States
e1: Control signal	a1: Start drive motor down	s1: Door up
e2: End of down track hit	a2: Start drive motor up	s2: Door down
e3: End of up track hit	a3: Stop drive motor	s3: Door stopped going down
e4: Laser beam crossed	a4: Reverse motor down to up	s4: Door stopped going up
		s5: Door closing
		s6: Door opening

13.3.3 Garage Door Controller VBA Code

```
' This is a simulator, not an actual software-controller hardware device.
' These constant variables correspond to the finite state machine model
Const e1 As String = ' control signal '
Const e2 As String = ' end of down track hit '
Const e3 As String = ' end of up track hit '
Const e4 As String = ' laser beam crossed '
Const a1 As String = ' start drive motor down '
```

```
Const a2 As String = ' start drive motor up '
Const a3 As String = ' stop drive motor'
Const a4 As String = ' reverse motor down to up '
Const s1 As String = ' Door Up '
Const s2 As String = ' Door Down '
Const s3 As String = ' Door stopped going down '
Const s4 As String = ' Door stopped going up '
Const s5 As String = ' Door closing '
Const s6 As String = ' Door opening
'

Dim inputEvent As String, outputAction As String, initialState, inState As String
Dim wantToContinue As Boolean
'

Output('Do you want to continue? Answer T or F)
Input(wantToContinue)
If (wantToContinue ) Then
        Output('enter initial state name e.g., s1')
        Input (initialState)
        inState = initialState
EndIf

While wantToContinue
        Select Case inState
                Case inState = s1   ' Door Up
                        Output('enter event name e.g., e1')
                        Input (inputEvent)
                        If e1
                                Then  Output('In state ', s1, 'Event ' ,e1, ' causes output ',a1, '.
                                        Next state is ', s5)
                                          inState = s5
                                Else  Output(inputEvent, ' cannot occur in state ', inState)
                        EndIf

                Case inState = s2  " Door Down
                        Output('enter event name e.g., e1')
                        Input (inputEvent)
                        If e1
                                Then  Output('In state ', s2 'Event ' ,e1 '
                                        causes output ',a2 '. Next state is ', s6)
                                          inState = s6
                                Else  Output(inputEvent, ' cannot occur in state ', inState)
                        EndIf

                Case inState = s3  ' Door stopped going down
                        Output('enter event name e.g., e1')
                        Input (inputEvent)
                                              Input (inputEvent)
                        If e1
```

Then Output('In state ', s3 'Event ' ,e1 ' causes output ',a1 '. Next
state is ', s5)
  inState = s5
Else Output(inputEvent, ' cannot occur in state ', inState)
EndIf

Case inState = s4 ' Door stopped going up
Output('enter event name e.g., e1')
Input (inputEvent)
If e1
Then Output('In state ', s4 'Event ' ,e1 ' causes output ',a2 '. next
state is ', s6)
  inState = s6
Else Output(inputEvent, ' cannot occur in state ', inState)
EndIf

Case inState = s5 ' Door closing
Output('enter event name e.g., e1')
Input (inputEvent)
Select Case inputEvent
  Case e1
    Output('In state ', s5, 'Event ' ,e1, ' causes output ',a3, 'next state
    is ', s3)
    inState = s3
  Case e2
    Output('In state ', s5, 'Event ',e2, ' causes output ',a3, 'next
    state is ', s2) inState = s2
  Case e4
    Output('In state ', s5, 'Event ',e4, ' causes output ',a4, 'next state is
    ', s6)
    inState = s6

  Else Output(inputEvent,' cannot occur in state ', inState)
End Select

Case inState = s6 ' Door opening
Output('enter event name e.g., e1')
Input (inputEvent)

Select Case inputEvent
  Case e1
    Output('In state ', s6, 'Event ',e1 ' causes output ',a3 '. next state is
    ', s4
    inState = s4

```
                        Case c3
                                Output('In state ', s6, 'Event ',e3 ' causes output ',a3 '. next
                                state is ', s1
                                inState = s1
                        Else Output('inputEvent cannot occur in state ', inState)
                End Select
        End Select

Output('Do you want to continue? Answer T or F)
Input(wantToContinue)
End While

End
```

Chapter 14

Smartesting Yest and CertifyIt

14.1 Introduction

Most of the material in this chapter was provided either by staff at Smartesting or taken (with permission) from the company website (www.smartesting.com). Smartesting has its headquarters in Besançon, France, and has offices in Paris. They can be contacted at hello@smartesting.com. Smartesting is a software product vendor and a strong player in model-based testing (MBT) with two products:

- *Yest*—a tool supporting test analysis, design, and implementation based on graphical models of business processes and business rules. Yest is dedicated to workflow testing of enterprise IT systems, enabling a smooth alignment between test cases and business requirements.
- *CertifyIt*—a general-purpose MBT tool, supporting strong capabilities to optimize the test coverage and industrialization of the test process (from requirements to manual or automated scripts). CertifyIt also supports capabilities to cover functional security requirements.

In addition to their two model-based testing products, they offer extensive consulting and training, including the training for ISTQB® Certified Model-Based Tester. One of the principals, Bruno Legeard, is a coauthor of *Practical Model-Based Testing*, [Utting 2007] and *Model-Based Testing Essentials—Guide to the ISTQB Certified Model-Based Tester: Foundation Level*, [Kramer 2016].

14.1.1 Smartesting Products

Smartesting has two main products, Yest for business applications and CertifyIt for more complex or event-driven systems. CertifyIt is based on the Smartesting family of test generation tools.

14.1.1.1 Yest

The Yest product is intended for enterprise IT (all business domains and large-scale IT information systems). It is a comprehensive solution that spans test analysis, test design, and test implementation. Based on a workflow analysis, Yest uses light-weight modeling of business processes—either formal business process models, such as Business Process Modeling and Notation (BPMN), or rule-based models that capture the business rules. This information is used to support test generation:

- *Test levels*: Acceptance and system test of enterprise IT systems
- *Testing types*: High-level functional and end-to-end testing
- *Typical users*: Testers, test analysts, functional testers, and business analysts

In Yest, the user may define business processes and/or test cases. A business process, representing tasks and alternatives, describes the dynamics of the application. Some business rules, depicted by decision tables or decision trees, are connected to the tasks of the process. Test cases can be generated by the tool, following a process and applying the related rules, or written by hand and compared to the process and rules. Once the test cases are acceptable to the user, they can be "published" into a test repository. Updating the model (process or rules) may update or invalidate the existing tests. Subsequent publications will update the test repository accordingly and increase the test designer's productivity. Yest is a stand-alone product, but multiple users can share and collaborate on Yest projects. For more detailed information, see the product website: http://yest.smartesting.com.

14.1.1.2 CertifyIt

The CertifyIt product is a general purpose MBT tool that uses behavioral modeling to automate test generation and execution, and industrialize the full test process from test objectives to test scripts. It is intended for embedded systems, complex systems, security critical systems, and API- or web service-based middleware applications such as ePayment, smart cards, the Internet of Things platforms, and cyber-physical systems.

CertifyIt supports both integration- and system-level testing with a choice of functional or robustness testing (including testing functional security requirements). Typical users include test designers, test analysts, test architects, and technical test analysts.

CertifyIt manages two views of the system under test:

- The static view of the system, made of a simplified representation of data structures, is modeled by UML class diagrams.
- The dynamic view of the system (its behavior) is modeled by scripts written in a behavioral language, a subpart of OCL (Object Constraint Language).

For more detailed information, see the product website: http://certifyit.smartesting.com.

A CertifyIt user can generate functional and robustness test cases that aim at testing the system's behavior. To generate functional test cases, CertifyIt automatically creates test targets based on a control flow analysis of the OCL expressions in the operations. It generates a set of test cases that covers each test target at least once.

CertifyIt also aims at robustness testing by applying a weighted exploratory testing approach or by generating tests to cover requirements for critical security functions in the systems. These security functional requirements are expressed by a test engineer using dedicated and user-friendly languages that easily express the expert's knowledge to produce high quality and security-dedicated test cases.

In the following sections, we present the two case studies of the book, respectively managed with Yest for the "Insurance Premium Calculation" example, and with CertifyIt for the Garage Door Controller.

14.1.2 Customer Support

The Smartesting Consulting team helps clients/users maximize the return on their investment and to drive the adoption of MBT into their organization. Smartesting consultants are trained and experienced in delivering focused solutions. They provide effective guidance as well as tactical delivery for a successful implementation of model-based testing in client organizations. The support includes:

- Coaching on modeling/CertifyIt model architecture
- Integration of MBT into the client's test process
- Knowledge transfer of model-based testing methodology
- Smartesting resources in resource augmentation model
- Legacy test repository optimization with Impulse solution
- Assistance with publishing into repositories and automation tools
- Connecting to requirement and business process management tools

14.2 Insurance Premium Results Using Yest

Yest is based on a graphical representation of business processes representing the business flows. The test cases are generated from the processes, or written by the user for specific needs and checked against the modeled processes. The corresponding business rules are associated with the tasks of the processes, using decision tables or decision trees.

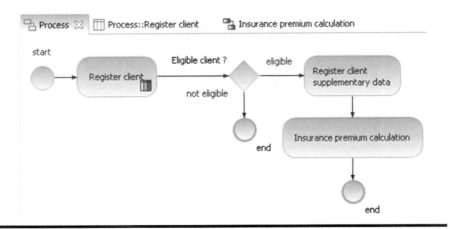

Figure 14.1 Yest process view for the Insurance Premium Calculation.

Smartesting personnel created a model of the "Insurance Premium Calculation" as a business process with three tasks in Figure 14.1. In the first task, the user enters the client's age and at fault claims in the Register client task to determine the client's eligibility. As per the problem statement, people younger than 16 or older than 90 are not eligible, as well as people having more than 10 "at fault" claims are not eligible. In the second task (Register client supplementary data), the information about students and drinking habits is requested in order to compute possible reductions for being a goodStudent and nonDrinker. The third task (Insurance Premium Calculation) computes the final insurance premium amount.

The Register client task uses the decision table in Figure 14.2 to determine coverage eligibility. This table is read from top to bottom. For an age older than 90 years, the second line applies with no consideration of the value of "At fault" claims, so the customer is not eligible.

The calculation process of the insurance premium is presented by a decision tree, as depicted in Figure 14.3. It models all calculation values needed to correctly apply the business rules. The decision tree complements the process view illustrated in Figure 14.3 and is associated with the task named "Insurance Premium Calculation" in the business process.

Yest generates test cases based on different test selection criteria (decision tree or table coverage) or by adding test scenarios manually. In this example, we generate test cases to cover the business rules defined in the calculation values as part of the business process and the decision table and tree. Based on the calculation values expressed in the decision tree, the tool automatically calculates the final amount of the insurance premium. If the client is eligible, the tool generates a test with two other steps where the supplementary data is requested for the Insurance Premium Calculation, and then the insurance premium is calculated.

The test cases can then be automatically published into a test repository, or to an excel spreadsheet. The final test cases produced are listed in Table 14.1.

Figure 14.2 Yest—eligibility decision table for the Insurance Premium Calculation.

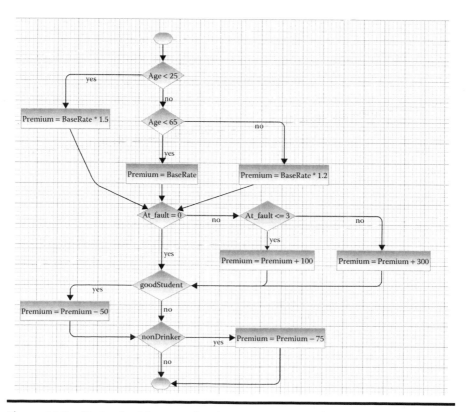

Figure 14.3 Yest—decision tree for the Insurance Premium Calculation.

Table 14.1   Concrete Test Cases Published by Yest

Test Case	Age	Claims	goodStudent	nonDrinker	Expected Premium
1	<16	Any	—	—	Not eligible
	Actions		**Expected results**		
	Enter client age: 14				
	Enter client at fault claims: 1				
	Validate entry		The client is not eligible		
2	>90	Any	—	—	Not eligible
	Actions		**Expected results**		
	Enter client age: 98				
	Enter client at fault claims: 5				
	Validate entry		The client is not eligible		
3	Any between 16 and 90	>10	—	—	Not eligible
	Actions		**Expected results**		
	Enter client age: 37				
	Enter client at fault claims: 18				
	Validate entry		The client is not eligible		

(Continued)

Table 14.1 (*Continued*) Concrete Test Cases Published by Yest

Test Case	Age	Claims	goodStudent	nonDrinker	Expected Premium
4	[16–25]	0	Yes	Yes	Premium = 775
	Actions		**Expected results**		
	Enter client age: 19				
	Enter client at fault claims: 0				
	Validate entry		The client is eligible		
	Enter client as a goodStudent				
	Enter client as a nonDrinker				
	Insurance Premium Calculation		The Insurance Premium Calculation is 775, and the offer is registered in the database		
5	[16–25]	0	No	Yes	Premium = 850
	Actions		**Expected results**		
	Enter client age: 21				
	Enter client at fault claims: 0				
	Validate entry		The client is eligible		
	Enter client data as not a goodStudent				
	Enter client as nonDrinker				
	Insurance Premium Calculation		The Insurance Premium Calculation is 850, and the offer is registered in the database		

(*Continued*)

Table 14.1 (Continued) Concrete Test Cases Published by Yest

Test Case	Age	Claims	goodStudent	nonDrinker	Expected Premium
6	[25–65]	1–3	No	No	Premium = 625
	Actions		**Expected results**		
	Enter client age: 43				
	Enter client at fault claims: 2				
	Validate entry		The client is eligible		
	Enter client data as not a goodStudent				
	Enter client as drinker				
	Insurance Premium Calculation		The Insurance Premium Calculation is 625, and the offer is registered in the database		
7	[65–90]	4–10	No	Yes	Premium = 945
	Actions		**Expected results**		
	Enter client age: 74				
	Enter client at fault claims: 6				
	Validate entry		The client is eligible		
	Enter client data as not a goodStudent				
	Enter client as nonDrinker				
	Insurance Premium Calculation		The Insurance Premium Calculation is 945, and the offer is registered in the database		

Notice that these are "concretized" test cases with actual values for the client's age, at fault claims, student, and nonDrinker status. The test cases in Table 14.1 are equivalent to a form of equivalence class testing (see [Jorgensen 2014]).

To conclude, Yest can generate as many test cases as desired by the user. Here the goal was to cover each line of the decision table and to cover each valuation in the decision tree. This set of test cases is extensible and mainly depends on the test objectives to be covered. The decision to cover one aspect or the other of the model is made by the user, commonly based on cost and risk mitigation.

14.3 Garage Door Controller Results Using CertifyIt

The CertifyIt tool accepts as input an MBT model, represented by a class diagram, to model the structure of the system under test, and an MBT behavioral modeling language description, such as simplified OCL expressions in the form of pre/postconditions of class operations to model the system's behavior. Before modeling begins, it is necessary to identify a set of appropriate (to the SUT) test objectives and test selection criteria. The quality of the generated test cases is largely determined by these early choices.

In the CertifyIt process, good practice includes building a Test Objectives Charter (TOC) for the identified test objectives. The TOC will represent the test objectives using two tags: REQ—a high-level requirement identifying the objective and AIM—a refinement of the requirement (see Table 14.3). The TOC for the Garage Door Controller is based on the table of events and states in Table 14.2 and the specification description in Chapter 13.

More concretely, the TOC of the Garage Door Controller, illustrated in Table 14.3, contains the four high-level input events (from Table 14.2) represented

Table 14.2 Garage Door Controller Events and States

Input Events	Output Events (Actions)	States
e1: Control signal	a1: Start drive motor down	s1: Door up
e2: End of down track reached	a2: Start drive motor up	s2: Door down
e3: End of up track reached	a3: Stop drive motor	s3: Door stopped going down
e4: Laser beam interruption	a4: Reverse motor down to up	s4: Door stopped going up
		s5: Door closing
		s6: Door opening

Table 14.3 Test Objectives Charter

@REQ		@AIM	
Control signal	**Down**	When door is up (state s1) (start drive motor down)	Recover going down (start drive motor down)
	Stop	When the door is closing (state s5) (stop drive motor)	Stop going down (stop drive motor)
	Down	When the door is stopped going down (state s3) (start drive motor down)	Recover going down (start drive motor down)
	Up	When door is down (state s2) (start drive motor up)	Recover going up (start drive motor up)
	Stop	When the door is opening (state s6) (stop drive motor)	Stop going up (stop drive motor)
	Up	When the door is stopped going up (state s4) (start drive motor up)	Recover going up (start drive motor up)
End of up track detection		End of up track reached (stop drive motor)	Unauthorized end of track detection event from states other than Door opening
End of down track detection		End of down track reached (stop drive motor)	Unauthorized end of track detection event from states other than Door closing
Light beam cross		Reverse motor (down to up)	Unauthorized light beam crossed event from states other than Door going down

as requirements (@REQ), based on the events analysis: control signal, end of down and up track detection, light beam interruption. The Control Signal event can occur in six different contexts (modeled here as states). Elsewhere, this is known as a context sensitive input event in which the same physical event results in different responses based on the context in which it occurs.

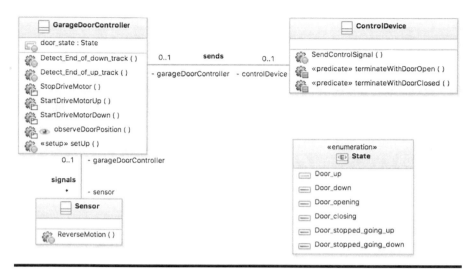

Figure 14.4   Garage Door Controller class diagram.

The required actions are the first three actions in Table 14.2, namely, start the motor down, start the motor up, and stop the motor. The @AIM section is used to specify the contexts.

As depicted in Figure 14.4, the structure of the system is modeled by three classes: GarageDoorController—driving the motor and thus controlling the door state, ControlDevice—sending controls, and Sensor—signaling interruption events to control the door motion. The door state is modeled by an enumeration State, which changes depending on the input events. The input events are modeled by class operations.

The door is closed if the end of down track is reached, or opened if the end of up track is reached; this is represented by the controller's operations Detect_End_of_up_track and Detect_End_of_down_track. The controller starts driving the motor up or down (operations StartDriveMotorUp and StartDriveMotorDown) or stops it (operation StopMotorDrive) with respect to the control signal sent by the control device operation SendControlSignal, respectively. The reverse motion down to up signal in case of a light beam cross is modeled by the operation ReverseMotion.

In addition, the corresponding actions of an event with respect to the given door state are represented by OCL pre/postconditions in the operations, interpreted as an action language. The preconditions define the criteria under which the actions can be activated and the postconditions are a set of actions, for instance: the output events and change of the door state. To illustrate the behavior modeling of an event, consider the event of sending a control signal to open a door—DoorUp, part of the operation SendControlSignal. The control signal event can open a door if the door is closed (State::Door_down) or if the door's motion has been interrupted by a control signal (State::Door_stopped_going_up), as depicted in Figure 14.5. As a result, the motor will start driving up, and the door state will be changed.

To keep the bidirectional traceability, the TOC can be imported into the tag browser, as depicted in Figure 14.6, and to avoid any manual error-prone rewriting, the tags can be easily dragged and dropped from the browser into the OCL postcondition.

```
---@REQ:control signal
if(self.garageDoorController.door_state = State::Door_down) then
    ---@AIM:when door is down
    ---@AIM:recover going up
    ---@AIM:start drive motor up
    self.garageDoorController.StartDriveMotorUp() and
    self.garageDoorController.door_state = State::Door_opening
else if (self.garageDoorController.door_state = State::Door_stopped_going_up) then
    ---@AIM:when door is stopped going up
    ---@AIM:recover going up
    ---@AIM:start drive motor up
    self.garageDoorController.StartDriveMotorUp() and
    self.garageDoorController.door_state = State::Door_opening
else
        . . .
endif
endif and
self.garageDoorController.observeDoorPosition(self.garageDoorController.door_state)
```

Figure 14.5   Garage Door Controller—Control Signal Event—DoorUp.

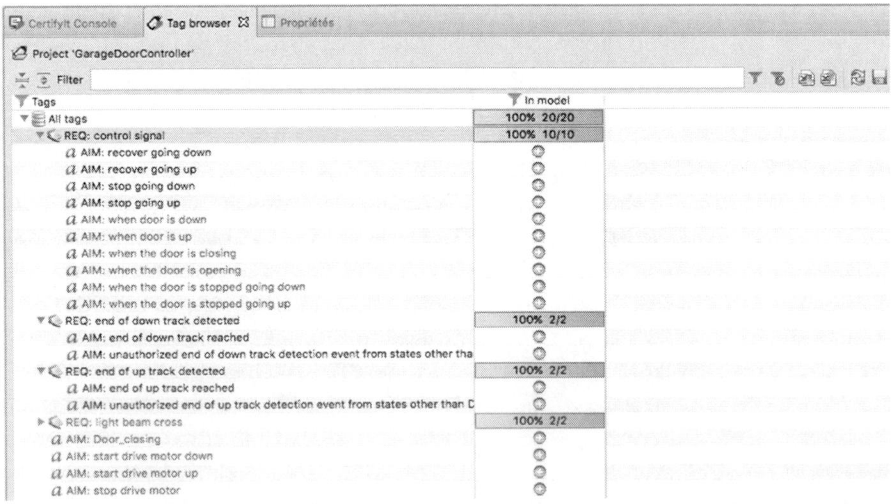

Figure 14.6   Certifylt Tag Browser.

To create functional test cases, the test generator defines test targets for each of the paths in the OCL expression (by analyzing the control flow of the expression). In this excerpt of the control signal event, two possible test targets can be distinguished: start opening the door when the door is closed (---@AIM:{when door is down, recover going up}) and recover the opening of the door when the door motion is stopped (--@AIM:{when door is stopped going up, recover going up}). Thus, the generator will create test cases to cover each test target at least once. In total, the CertifyIt tool creates the following 12 test targets that will be covered by at least one generated test for each garage door initial state (see Figure 14.7).

We represent these two initial states (door opened and door closed) and the test data used for test generation as object diagrams, which is depicted in Figure 14.8.

Model element	∂ Aims	◔ Requirements
ControlDevice::SendControlSignal()	control signal/when door is down, control signal/recover going up	control signal
ControlDevice::SendControlSignal()	control signal/when the door is stopped going up, control signal/recover going up	control signal
ControlDevice::SendControlSignal()	control signal/when door is up, control signal/recover going down	control signal
ControlDevice::SendControlSignal()	control signal/when the door is stopped going down, control signal/recover goin...	control signal
ControlDevice::SendControlSignal()	control signal/when the door is opening, control signal/stop going up	control signal
ControlDevice::SendControlSignal()	control signal/when the door is closing, control signal/stop going down	control signal
GarageDoorController::Detect_End_...	end of down track detected/end of down track reached	end of down track detected
GarageDoorController::Detect_End_...	end of down track detected/unauthorized end of down track detection event fro...	end of down track detected
GarageDoorController::Detect_End_...	end of up track detected/end of up track reached	end of up track detected
GarageDoorController::Detect_End_...	end of up track detected/unauthorized end of up track detection event from sta...	end of up track detected
Sensor::ReverseMotion()	light beam cross/reverse motor down to up	light beam cross
Sensor::ReverseMotion()	light beam cross/unauthorized light beam crossed event from states other than...	light beam cross

Figure 14.7   Garage Door Controller—CertifyIt Test Targets.

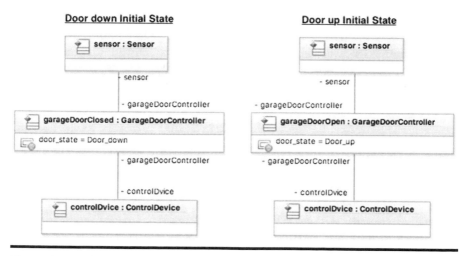

Figure 14.8   Garage Door Controller test data.

To automate the test oracle in the MBT model, an observing operation is added (observeDoorPosition) that will construct an oracle at each test step to validate the current door state. Thus, if the returned system state differs from the expected state, the test will fail. Finally, in order to reset the system state at initial state, two predicate functions that are used only for test generation are added: terminateWith-DoorClosed and terminateWithDoorOpen (see Figure 14.4), with respect to the initial state of the system (door closed or opened).

To generate test cases with each of the initial states, for illustration purpose, two separate test suites were created.

Then the test generator based on the available test data and the identified test targets will generate tests automatically to cover each of the test targets at least once, which correspond to the objectives defined in the TOC.

Consider, for instance, the initial state of closed door and the Control Signal Event—DoorUp depicted in Figure 14.5. The test generator will create two test targets that will be covered by one test case (test case 1), illustrated in Table 14.4. In addition, as given in the table, CertifyIt tool generates 7 tests to cover the 12-test targets for the initial state—door is down.

Figure 14.9 illustrates the abstract test cases for the two test suites in the CertifyIt test generator. The test case 1 can be seen at the level of the last observation in the *Test detail* panel on the right side of the figure.

These abstract test cases can then be directly published into JUnit test cases, or for documentation purposes into a web page or any other format or test management tool. Here it is exported into TestLink for further test management. Figure 14.10 illustrates the test case 1 exported into JUnit, executable on the Garage Door Java implementation using a concretization layer.

The concretization layer maps the abstract values to the concrete ones in the system. Figure 14.11 shows the adaptor for the Control Signal Event—DoorUp. For instance, the step calls the abstract function SendControlSignal, which activates the concrete control signal event to the Garage Door Controller, through the function getConcreteControlSignalValue (as shown on the right part of the Figure 14.11).

To conclude, the 14 functional tests were executed on the Garage Door Controller Java implementation and all of them passed successfully.

In addition, CertifyIt offers two other extensions to generate test cases dedicated to functional security testing, based on temporal properties and test purposes. The temporal properties express properties of the systems in terms of successions, precedence, or encapsulation of events, whereas the test purposes express procedures that may exercise specific corner cases of the system, based on the expert's knowledge. These extensions complement the functional testing achieved by test targets coverage.

Table 14.4 Generated Test Cases from CertifyIt MBT Model, Initial State = State Down

Test Case	Test Steps	In State	Input Events	Output Events (Actions)	Next State
#				*Generated Test Cases*	
1	Step 1	s2: Door down	e1: Control signal	a1: Start drive motor up	s6: Door opening
	Step 2	s6: Door opening	e1: Control signal	a3: Stop drive motor	s4: Door stopped going up
	Step 3	s4: Door stopped going up	e1: Control signal	a1: Start drive motor up	s6: Door opening
2	Step 1	s2: Door down	e1: Control signal	a1: Start drive motor up	s6: Door opening
	Step 2	s6: Door opening	e3: End of up track hit	a3: Stop drive motor	s1: Door up
	Step 3	s1: Door up	e1: Control signal	a1: Start drive motor down	s5: Door closing
	Step 4	s5: Door closing	e2: End of down track hit	a3: Stop drive motor	s2: Door down
3	Step 1	s2: Door down	e1: Control signal	a1: Start drive motor up	s6: Door opening
	Step 2	s6: Door opening	e3: End of up track hit	a3: Stop drive motor	s1: Door up
	Step 3	s1: Door up	e1: Control signal	a1: Start drive motor down	s5: Door closing
	Step 4	s5: Door closing	e1: Control signal	a3: Stop drive motor	s3: Door stopped going down

(Continued)

Table 14.4 (*Continued*)   Generated Test Cases from CertifyIt MBT Model, Initial State = State Down

Generated Test Cases

Test Case		In State	Input Events	Output Events (Actions)	Next State
	Step 5	s3: Door stopped going gown	e1: Control signal	a1: Start drive motor down	s5: Door closing
4	Step 1	s2: Door down	e1: Control signal	a1: Start drive motor up	s6: Door opening
	Step 2	s6: Door opening	e3: End of up track hit	a3: Stop drive motor	s1: Door up
	Step 3	s1: Door up	e1: Control signal	a1: Start drive motor down	s5: Door closing
	Step 4	s5: Door closing	e4: Laser beam interruption	a4: Start drive motor up	s6: Door opening
5	Step 1	s5: Door down	e2: End of down track hit	Unauthorized	s2: Door down
6	Step 1	s6: Door up	e3: End of up track hit	Unauthorized	s1: Door up
7	Step 1	s5: Door down	e4: Laser beam interruption	Unauthorized	s6: Door down

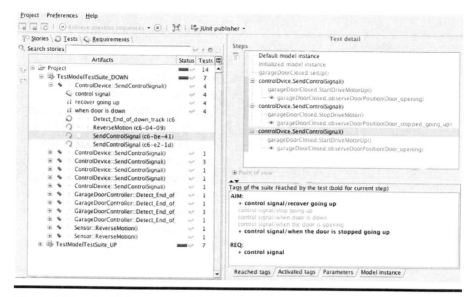

Figure 14.9 **CertifyIt generated test cases.**

```
/*
REQUIREMENTS:
  control signal
*/
public class SendControlSignal_c6_be_41_ extends TestCase {

  private AdapterImplementation adapter;

  public void setUp() throws Exception {
    adapter = new Adapter();
    adapter.setUp(garageDoorClosed);
  }

  public void testSendControlSignal__c6_be_41_() throws Exception {
    adapter.SendControlSignal(controlDevice);
    adapter.checkState(garageDoorClosed, Door_opening);
    adapter.SendControlSignal(controlDevice);
    adapter.checkState(garageDoorClosed, Door_stopped_going_up);
    adapter.SendControlSignal(controlDevice);
    adapter.observeDoorPosition(garageDoorClosed, Door_opening);
  }

  public void tearDown() throws Exception {
    adapter.closeAdapter();
  }

}
```

Figure 14.10 **Garage Door Controller—JUnit export of test case 1.**

```
@Override
public void SendControlSignal(ControlDevice device){
  currentState = getConcreteControlSignalValue(device);
}
```

```
EventWatcher.setCurrentEvent(ControlSignal.e1_control_signal)
GarageDoor.simulate();
return GarageDoor.getDoorState();
```

```
@Override
public void observeDoorPosition(GarageDoorController controller,
                 State expected_State) {
  Assert.assertTrue(currentState ==
              getConcreteStateValue(expected_State));
}
```

```
case Door_closing:
      return GarageDoorState.Door_closing;
case Door_down:
      return GarageDoorState.Door_down;
case Door_opening:
      return GarageDoorState.Door_opening;
case Door_stopped_going_down:
      return GarageDoorState.Door_stopped_going_down;
case Door_stopped_going_up:
      return GarageDoorState.Door_stopped_going_up;
case Door_up:
      return GarageDoorState.Door_up;
```

Figure 14.11 **Garage Door Controller adaptation layer (for DoorUp).**

14.4 Vendor Advice: Best Practices with Yest and CertifyIt

This section puts in a nutshell the Smartesting advice with respect to its two products: Yest and CertifyIt. These tools support different target audiences; they address different test objectives to be covered and thus require different skills.

Yest is dedicated to functional testers during test analysis, test design, and test implementation phases. It starts from lightweight modeling of business processes and rules, using simple graphical notations (business process modeling, decision tables, and decision trees). Yest can be used with an iterative approach. The user may start from initial modeling, and/or from test cases, and then align the artifacts: test objectives, business processes, business rules, and test cases. This iterative approach will maximize the efficiency of the test analysis and the design and implementation activities supported by Yest.

CertifyIt requires some technical skills, so the most effective way of using the tool is to have dedicated users, just like for test automation. Introducing CertifyIt into the existing testing process also means introducing the tool in an existing tool chain. So for a smooth integration, CertifyIt must also be customized (mainly publishers). We recommend getting Smartesting involved in this process to get a quick and efficient return on investment.

References

[Jorgensen 2014]
Jorgensen, Paul, *Software Testing: A Craftsman's Approach, fourth edition*. CRC Press, Taylor and Francis Group, Boca Raton, Florida, 2014. ISBN 978-1-4665-6068-0.
[Kramer 2016]
Kramer, Anne and Legeard, *Bruno, Model-Based Testing Essentials-Guide to the ISTQB® Certified Model-Based Tester Foundation Level*, John Wiley and Sons, Hoboken, New Jersey, 2016. ISBN 9781119130017.
[Utting 2007]
Utting, Mark and Legeard, Bruno, *Practical Model-Based Testing: A tools Approach*, Morgan Kaufmann, San Francisco, 2007. ISBN-10: 0 12-372501-1.

Chapter 15

TestOptimal

15.1 Introduction

Most of the material in this chapter was provided either by staff at TestOptimal, or taken (with permission) from the company website. TestOptimal LLC is located near Rochester, Minnesota, the United States. The product line is very comprehensive. The company website (http://testoptimal.com/) lists the following features:

- MBT Modeling—(Extended) Finite State Machine (EFSM)/State Diagram (UML), and Control Flow Graph (CFG)/Activity Diagram (UML)
- SuperState and SubModel—organizes and partitions a larger model into smaller reusable library components
- Graphs—model graph, sequence graph, coverage graph, and message sequence chart (MSC)
- Model import/Merge—UML XMI model and other XML-based graph modeling formats (GraphXML and GraphML)
- Test Case Generation—random walk, optimal sequencer, mutant path, priority path, and mCase (custom test case) sequencer
- Scripting—mScript (XML-based scripting)
- Data-Driven Testing (DDT)—embeds data-driven testing in the model and uses external data source to generate test cases
- Behavior-Driven Testing (BDT)—BDT style tests design and tests automation with upgrade path to MBT (beta)
- JDBC/ODBC support—provides access to relational databases to read, write, store, and verify test result
- WebSvc/RESTful—tests RESTful web services

- Integration—ALM, Java IDE (Eclipse, NetBeans), JUnit, batch/cron, REST websvc, and remote agent to integrate with other test automation tools (such as QTP)
- Cross Browser—tests web applications on IE, Firefox, Chrome, Opera, and Safari
- Extensibility—customizes plug-in capability to test various types of applications
- Debug—sets break points, steps through model execution, visually highlights during debugging, automatically logs test steps that lead to the failures
- Load Testing—virtual users and realistic simulation of production load
- Dashboard/KPI—detailed and summarized test/requirement coverage, failures and performance statistics, charts, and reports. Configurable Key Performance Indicator (KPI)
- IDE Web App—browser-based application and mobile client
- Security—Ldap- and file-based HTTP authentication to prevent unauthorized access
- Requirement Traceability—tags requirements to states, transitions, or mSript
- Test Data Generation—various ways to generate test data including pairwise, combinatorial algorithms, and pattern-based data generation
- Model Animation—visualization of transition traversals on the model
- Test Multiple Types of Applications Simultaneously—synchronizes testing of web application, windows application, and backend process at the same time

15.1.1 Product Architecture

The TestOptimal features are carefully integrated into the architecture as shown in Figure 15.1.

15.1.2 The TestOptimal Product Suite

The TestOptimal Suite is a package of MBT tools for both functional (specification-based) and load/performance testing. It includes the following products:

- BasicMBT—supports modeling and test case generation for either manual or offline testing. Generated test cases are saved to either text files or Excel spreadsheets.
- ProMBT—supports modeling and test automation for web-based and user interface applications. Both data-driven testing and combinatorial testing are supported, with connection to a background database. There is also the possibility of rule-based test data design to optimize testing efficiency.
- EnterpriseMBT—supports concurrent system modeling and automated load and performance testing.

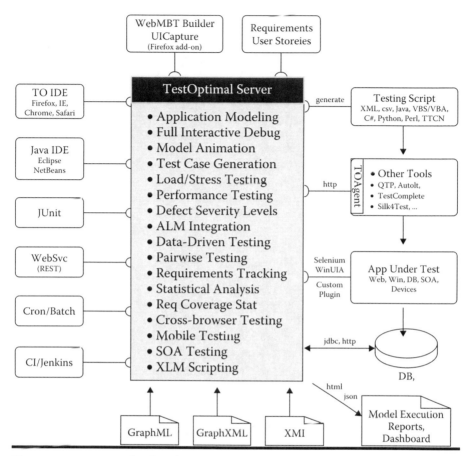

Figure 15.1 TestOptimal architecture Overview (5.1).

■ DataDesigner—offers rule-based design of combinatorial test data using the well-known orthogonal array algorithm. Supports pairwise, three-wise, ..., *n*-wise test case generation.

■ Behavior-Driven Testing/Behavior-Driven Development (BDT/BDD)—creates features and test scenarios in BDT/BDD style: GIVEN/WHEN/THEN. Combined with the power of MBT to design the efficient test automation suite for web, windows UI applications, and non-UI processes/interfaces.

■ Test data designer—aggregates and summarizes test execution, test coverage, and requirement traceability in Dashboard charts and reports. Configurable Key Performance Indicator (KPI) can be subscribed by individual test engineer's IDE.

■ RuntimeMBT—Runtime servers are used as dedicated QA servers to run test models and for a large-scale load and performance testing.

■ SvrMgr—SvrMgr is used to manage Runtime servers, acting as a central repository for models and test execution results. SvrMgr is responsible for distributing model updates to all Runtime servers and dispatch models to run on appropriate Runtime servers.

15.1.3 Customer Support

The TestOptimal website provides five tutorials and three supplemental presentations. In addition, there is an online form to submit questions directly to TestOptimal staff and a user forum, where TestOptimal users can ask questions and share their experiences and best practices using TestOptimal.

15.2 Insurance Premium Results

The DataDesigner is a rule-based test case design tool that uses six variations of the orthogonal array algorithm to generate a minimum number of test cases for a given set of test variables. For the Insurance Premium Problem, test cases for pairwise, three-wise, and four-wise outputs were generated, but due to the large number of test cases, only the pairwise test cases are shown in Table 15.1. The three-wise algorithm generated 18 test cases, and the four-wise algorithm generated 36 test cases. As there are four independent variables, the 36 four-wise test cases traverse each of the 36 rules in the decision table definition. In all three sets of results, the n-wise domains were as follows:

Age: 16–24, 25–64, 65–89	goodStudent: true, false
AtFaultClaim: 0, 1–3, 4–10	nonDrinker: true, false

Figure 15.2 shows the DataSet Definition in TestDesigner. The four columns in the Field Definition section represent four test variables and their domains as described earlier. The DataTable section shows the abstract test cases generated by the selected pairwise algorithm. The concrete test cases are shown in Figure 15.3.

The TestOptimal DataDesigner is also capable of filtering out certain permutations using user-specified rules to produce test cases with various coverage for a subset of fields. This offers a very powerful way to generate test cases for complex test variables, some of which may have more interaction among themselves (various interaction strength). Running the DataDesigner for higher levels of variable interaction naturally results in more test cases. The nine pairwise test cases expand to 18 with the three-wise option and expand to 36 with the four-wise option.

Table 15.1   Pairwise Abstract Test Cases (Full Set)

Test Case	Age	Claims	goodStudent	nonDrinker	Expected Premium
1	16–24	0	False	False	775
2	16–24	1–3	True	False	1000
3	16–24	4–10	False	True	1150
4	25–64	0	True	False	525
5	25–64	1–3	False	True	575
6	25–64	4–10	True	False	900
7	65–89	0	False	True	770
8	65–89	1–3	True	False	1125
9	65–89	4–10	False	False	970

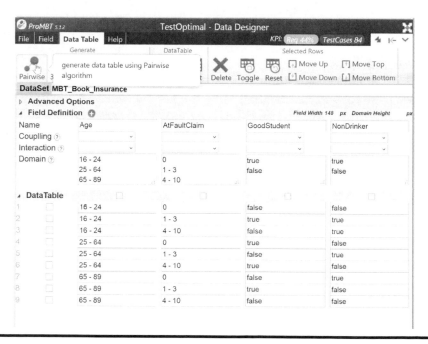

Figure 15.2   DataSet definition—Abstract pairwise test cases.

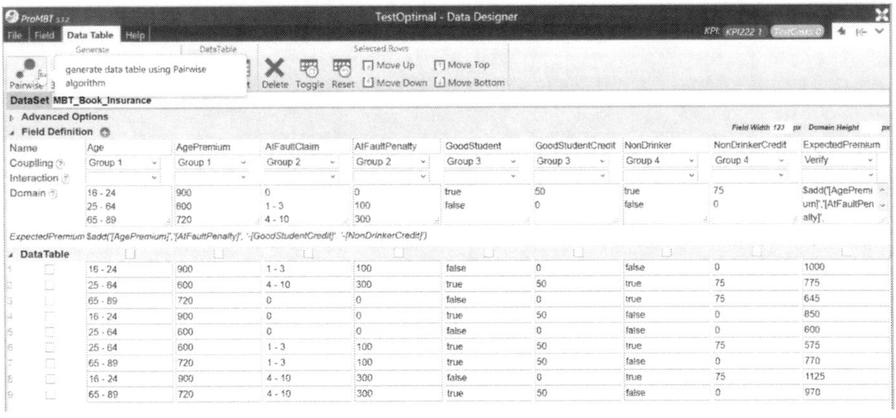
TestOptimal - Data Designer
generate data table using Pairwise algorithm
Delete Toggle Reset Move Up Move Down Move Top Move Bottom
DataSet MBT_Book_Insurance
Advanced Options
Field Definition
Name Age AgePremium AtFaultClaim AtFaultPenalty GoodStudent GoodStudentCredit NonDrinker NonDrinkerCredit ExpectedPremium
Coupling Group 1 Group 1 Group 2 Group 2 Group 3 Group 3 Group 4 Group 4 Verify
Interaction
Domain 16 - 24 / 25 - 64 / 65 - 89 | 900 / 600 / 720 | 0 / 1 - 3 / 4 - 10 | 0 / 100 / 300 | true / false | 50 / 0 | true / false | 75 / 0 | $add('[AgePremium]','[AtFaultPenalty]',
ExpectedPremium $add('[AgePremium]','[AtFaultPenalty]', '-[GoodStudentCredit]' '-[NonDrinkerCredit]')
DataTable
1 16 - 24 900 1 - 3 100 false 0 false 0 1000
2 25 - 64 600 4 - 10 300 true 50 true 75 775
3 65 - 89 720 0 0 false 0 true 75 645
4 16 - 24 900 0 0 true 50 false 0 850
5 25 - 64 600 0 0 false 0 false 0 600
6 25 - 64 600 1 - 3 100 true 50 true 75 575
7 65 - 89 720 1 - 3 100 true 50 false 0 770
8 16 - 24 900 4 - 10 300 false 0 true 75 1125
9 65 - 89 720 4 - 10 300 true 50 false 0 970

Figure 15.3 DataSet definition—Concrete pairwise test cases.

15.3 Garage Door Controller Results

Since the Garage Door Controller is an event-driven system (with no numerical calculation), the most a tool can do is to generate paths that are easily turned into test cases. Since finite state machines are the most popular choice for event-driven systems, each vendor was provided with the same model. It included the states and events that are shown in Table 15.2.

The TestOptimal ProMBT graphic modeling IDE was used to create the FSM model for the Garage Door Controller problem as shown in Figure 15.4. The nodes represent the state of the Garage Door with initial state set to DoorUp. From this model, the test cases are generated by the sequencer.

Test cases were generated by Optimal Sequencer, one of seven sequencers that generate test cases/sequences from the model. Optimal Sequencer generates the shortest test sequence (steps) that covers 100% of transitions in the model. Each of the sequencers generates a set of test cases to achieve different test objectives and test coverage.

The TestOptimal ProMBT tool produces several useful forms of output for the Garage Door Controller—a UML-style sequence chart (see Figure 15.5), a directed graph showing test sequence coverage of states in the finite state machine model (Figure 15.6), and a set of test cases driven by permissible event sequences (see Table 15.3).

TestOptimal ProMBT also provides extensive test coverage information. Figure 15.6 is a directed graph of the generated test sequences. In the test sequence graph, nodes are the given states of the Garage Door Controller finite state machine, and edges correspond to the control events. As should be expected, TestOptimal

Table 15.2 Garage Door Controller Events and States

Input Events	Output Events (Actions)	States
e1: Control signal	a1: Start drive motor down	s1: Door up
e2: End of down track hit	a2: Start drive motor up	s2: Door down
e3: End of up track hit	a3: Stop drive motor	s3: Door stopped going down
e4: Laser beam crossed	a4: Reverse motor down to up	s4: Door stopped going up
		s5: Door closing
		s6: Door opening

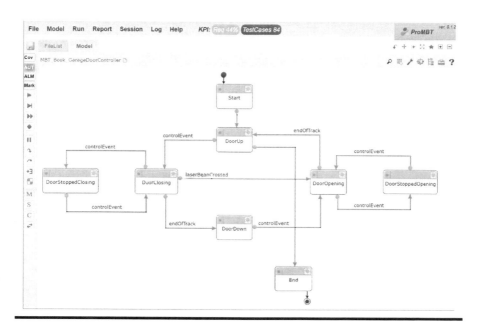

Figure 15.4 Garage Controller FSM in TestOptimal IDE.

produces the necessary test coverage reports. Table 15.4 is the test coverage report for the test case in Table 15.3.

It is possible to run the model with different sequencers to produce test cases that achieve different test coverage. The Optimal sequencer produces the sequence (shortest test steps) with minimum number of test steps while

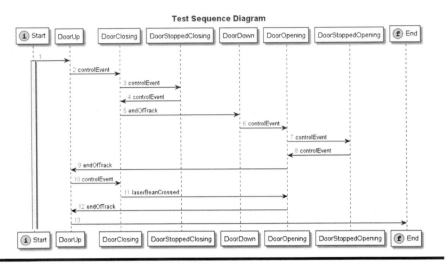

Figure 15.5 Sample sequence diagram for the Garage Door Controller.

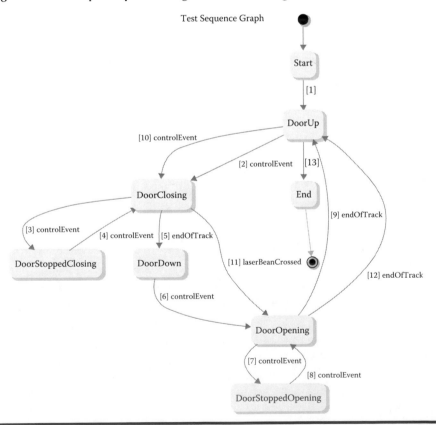

Figure 15.6 State coverage of test cases.

Table 15.3 Sample Test Case

TestCase ID	Step	Action	Expected Results	Actual Results
0001	(Weight: 55, length: 11)			
	1	From: DoorUp Do: controlEvent Input: controlEvent	At: DoorClosing Expect Garage Door State: DoorClosing	
	2	From: DoorClosing Do: controlEvent Input: controlEvent	At: DoorStoppedClosing Expect Garage Door State: DoorStoppedClosing	

	3	From:
		  DoorStoppedClosing
		Do: controlEvent
		Input: controlEvent
			At: DoorClosing
			Expect Garage Door
			  State: DoorClosing	

	4	From: DoorClosing Do: endOfTrack Input: endOfTrack	At: DoorDown Expect Garage Door State: DoorDown	
	5	From: DoorDown Do: controlEvent Input: controlEvent	At: DoorOpening Expect Garage Door State: DoorOpening	

	6	From: DoorOpening
		Do: controlEvent
		Input: controlEvent
			At:
			  DoorStoppedOpening
			Expect Garage Door
			  State:
			  DoorStoppedOpening	

	7	From:
		  DoorStoppedOpening
		Do: controlEvent
		Input: controlEvent
			At: DoorOpening
			Expect Garage Door
			  State: DoorOpening	

	8	From: DoorOpening Do: endOfTrack Input: endOfTrack	At: DoorUp Expect Garage Door State: DoorUp	
	9	From: DoorUp Do: controlEvent Input: controlEvent	At: DoorClosing Expect Garage Door State: DoorClosing	

(Continued)

Table 15.3 (*Continued*) Sample Test Case

TestCase ID	Step	Action	Expected Results	Actual Results
	10	From: DoorClosing Do: laserBeanCrossed Input: laserBeanCrossed	At: DoorOpening Expect Garage Door State: DoorOpening	
	11	From: DoorOpening Do: endOfTrack Input: endOfTrack	At: DoorUp Expect Garage Door State: DoorUp	

Table 15.4 Test Coverage Report

Model Name	MBT_Book_Garage Door Controller
Model Version	1.0
Model Description	
Sequencer	OptimalSequence
Generation Date	2016-02-28
Coverage Type	Alltrans
Transition Coverage	100
Test Case Count	1
AUT Version	
TO Version	5.1.2
Requirements Version	
Stop Conditions	
Duplicate TestCases Removed	0

achieving 100% coverage of the model (all transitions covered). The PathFinder sequencer can be used with different parameters to produce various test coverage of different permutations and variations of paths. We have included the test cases generated in separate excel spreadsheets as well as the traversal graph and test case MSC charts to help visualize the test cases. TestOptimal supports several other sequencers as well as requirement tracking and requirement coverage as stopping criteria.

15.4 Vendor Advice

We have shown how TestOptimal generates test cases for offline MBT. TestOptimal also supports online MBT which is well suited for functional testing and load/performance testing of web/UI applications, web services, and business workflows.

Organizations considering adopting TestOptimal (or any MBT tools) into their testing/QA process should consider trying it on a small/pilot project first. Through the pilot project, an organization can learn and adjust its testing process and testing practice to make the best of the tools. Things to consider include how the tool will be integrated with the existing software development process and systems that are currently used to manage the testing process. Ask whether a product provides test reports for testers and management at different levels, and whether it has the development environment to help debug both the model and the automation test suite.

Chapter 16

Conformiq, Inc.

16.1 Introduction

Most of the material in this chapter was provided either by staff at Conformiq, or taken (with permission) from the company website. Conformiq has its headquarters in San Jose (California), main R&D center in Helsinki (Finland), with branch offices in Stockholm (Sweden), Bangalore (India), and Munich (Germany). Their product line, Conformiq 360° Test Automation, is very comprehensive and goes far beyond test generation, by integrating with existing software development life cycle (SDLC) tools in the testing process starting from requirements management and application lifecycle management (ALM) through test management and documentation, and automatic test execution tools. Visit their company website (https://www.conformiq.com/) for a list of mainline products and services.

16.1.1 Features

The Conformiq product line is extremely comprehensive, including capabilities for test case generation, two modeling languages, a user interface, and a flexible interface with other tools.

16.1.1.1 Test Generation

- Automatic generation of test inputs, including basic and structural data
- Automatic generation of expected test outputs, including basic and structural data
- Automatic generation of test timers
- Automatic support for requirements-driven test generation and requirement traceability

- Automatic support for boundary value analysis, multiple condition/multiple decision coverage, 2-transitions, all paths, combinatorial data, and many other black-box test design heuristics
- Automatic support for risk-based test generation
- Automatic generation of human-readable HTML or Excel-based test documentation in different target languages
- Automatic generation of executable test suites
- Automatic dependency ordering between generated test cases
- Automatic identification of test preambles, test bodies, and test postambles
- Automatic mathematical optimization of the generated test suites for efficiency and minimal test steps
- Automatic generation of traceability information
- Automatic generation of test case dependency information
- Automatic generation of test case names and test case summary
- Fully deterministic test generation based on constraint solving and symbolic state space exploration
- Incremental test generation and automatic test asset analysis
- Ability to distribute and scale test generation across multiple cores or on the cloud

16.1.1.2 Modeling Language (Conformiq Designer)

- Support for hierarchical UML2 state machine diagrams
- Support for direct importing state machine diagrams (and other diagram types such as class diagrams) from third-party UML tools for test generation
- Java-compatible textual notation as UML action language
- Full support for object-orientation, including classes, inheritance, multiple threads, and asynchronous communication between model threads
- Support for modeling multiple testing interfaces
- Support for timers
- Arbitrary-precision arithmetic on model level

16.1.1.3 Modeling Language (Conformiq Creator)

- Fully graphical modeling notation designed for testers that requires no background in programming
- Domain specific specification of interfaces in structure diagrams with direct support for modeling user interfaces and message-based interfaces with structured data and allowing extension via custom actions
- Automatic generation of action keyword repository and data objects from structure diagrams
- Support for user-defined structure diagram libraries
- Special actions for requirement annotation, narratives, and for influencing test name generation

- Support for hierarchical specification of activity diagrams, including control and data flows as well as complex and compound conditions
- Special concept of "don't care" test data value and direct support for multiple types of test data object (variable, value, alternative values, and value list)
- Support for modeling test data in spreadsheets
- Live checks of diagrams for syntactical correctness and quick fixes during editing, enabling change impact analysis

16.1.1.4 User Interface

- Fully interactive environment for modeling and review of generated tests built on top of Eclipse®
- Support for multiple test target settings per project
- Interactive graphical browsing of the generated test cases in message sequence diagrams with data in hierarchical lists as well as animated diagrams
- Interactive test target view with test traceability information
- Interactive test case dependency matrix view
- One-click generation of test cases
- One-click export of test cases in any supported format
- Built in Conformiq modeling editors designed for modeling for testing

16.1.1.5 Integration with Other Tools (see Figure 16.1)

- Fully user-definable output formats based on a open, Java-based scripting backend plugin architecture
- Import UML models from IBM Rhapsody, Sparx Enterprise Architect, IBM RSD-RT
- Create Creator models from manual tests specified in Excel, Gherkin feature files, or imported from flowcharts specified in MS Visio or BPMN tools such as Software AG Aris, IBM RSA, Mega, etc.
- Create structure diagrams from HP QTP/UFT or Selenium test automation assets
- Download requirements for model annotation automatically from requirement management tools such as HP Quality Center, HP ALM, IBM DOORS, IBM RequisitePro, Rally, MS Excel, or company-proprietary ones
- Publish test cases automatically to test management tools such as HP Quality Center, HP ALM, Rally, or company-proprietary ones
- Publish tests for automatic test execution either via Conformiq Transformer or directly to HP QTP/UFT, Selenium, Parasoft SOATest, Tricentis Tosca, Odintech Axe, Testplant Eggplant, Experitest SeeTest, Robot Testing Framework, etc., as well as any JUnit/Java, Python, TCL, XML, Perl, TTCN-3, etc., based testing framework
- Install Conformiq as an Eclipse plugin with other tools inside the Eclipse workbench
- Graphical compare and merge with GIT versioning control, but also compatible with other version management and SCM tools

16.1.1.6 Other Features

- Cross-platform solution with support for both Windows and Linux installation
- Deployable with local and remote test generation
- User manual and example models
- Quick start program
- Floating licenses and named-user licenses available

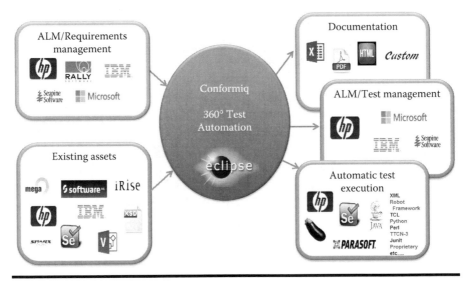

Figure 16.1   Conformiq architecture overview.

16.1.2 The Conformiq 360° Test Automation Product Suite

The Conformiq **360° Test Automation** product line contains five tools, which are briefly described in this section. The Insurance Premium Problem was done with Conformiq Creator; the Garage Door Controller was done with Conformiq Designer.

Conformiq Creator automatically tests enterprise IT applications, web applications, and web services. Creator is designed to automate functional, application, system, and end-to-end testing. This product requires no programming experience and supports fully graphical modeling for testing.

Conformiq Designer uses the power of Java in combination with UML2, which allows for the simple test automation of complex systems, such as embedded software and network equipment. Highly intelligent algorithms automatically generate the necessary tests, which aids in improving software quality.

Conformiq Transformer automates the execution of user-specified and automatically generated tests, supporting a wide range of industry-standard SDLC tools, and improving the productivity of the entire quality assurance organization. Working with Conformiq Creator, it extends the Conformiq 360° Test Automation solution.

Conformiq 360° Integrations offers important enablers that allow for easy integration with Application Lifecycle Management and other systems in Software Development Life Cycle with Conformiq Designer and Conformiq Creator.

Conformiq Grid is a companion product for Conformiq Designer and Conformiq Creator. It is a scalable cloud solution for test generation, which allows for organizations to effectively utilize Conformiq products and add more capacity when necessary.

16.1.3 Customer Support

Conformiq has an interesting process for deployment of its product line with new customers. For more details, please see https://www.conformiq.com/deployment/. For our purposes, the steps in the Conformiq deployment process are:

1. An initial technical advisory period to help customers develop a successful plan.
2. An optional proof of concept plan customized to customer needs.
3. A pilot project that is scoped together with the customer in such a way that allows a quick assessment of the return on investment from using Conformiq technology.
4. Advice on how to integrate Conformiq products into a customer-defined software process.
5. Using the Conformiq 360° Integrations product to move the Conformiq product(s) smoothly into a comprehensive ALM system.
6. Introduction and training classes.
7. Deployment with a "center of excellence" staffed by in-house champions of the model-based testing (MBT) process.
8. Continued technical development with workshops, skill assessment, and coaching.
9. Frequent model and generated test reviews.

These initial steps can be complemented by further customized training, support, and other follow-up possibilities.

16.2 Insurance Premium Results

The Conformiq process begins with using Conformiq Creator to develop the activity diagram in Figure 16.2. The Creator modeling language is intentionally limited to strictly binary decisions to simplify modeling and review of models; it is an equivalent rewrite of the one provided to all MBT tool vendors.

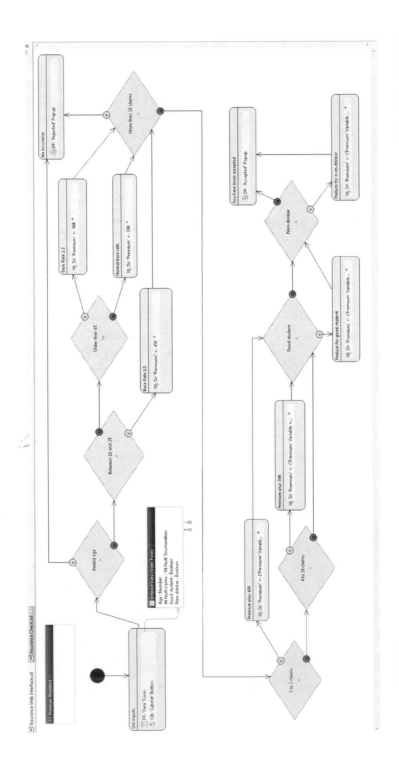

Figure 16.2 Activity diagram for the insurance premium problem.

16.2.1 Conformiq Creator Input

The text in the decision boxes, from left to right, and then top to bottom is as follows:

1. Invalid age?
2. Between 16 and 25?
3. Older than 65?
4. More than 10 claims? (then in the lower part ...)
5. 1 to 3 claims?
6. 4 to 10 claims?
7. goodStudent?
8. nonDrinker?

The actions shown as activities, that is, the rounded rectangles, are used to model the interactions via the interfaces with the system or application under test as well as computations needed to compute the premium and its applicability. The structure diagram in Figure 16.3 defines the interfaces available for testing; it is the basis from which the actions that interact via these interfaces are automatically generated. In this example, we modeled the interface as a web application user interface where the user can enter data via a form on a screen and then gets either a popup with a label displaying the premium showing the acceptance or otherwise popup showing the rejection after clicking the submit button on the previous screen. The diagram also illustrates different options for modeling underlying data types—in the example, the age field is modeled as a number, whereas the "at fault claims" are modeled as an enumeration, and "goodStudent" and "nonDrinker" are modeled using a (Boolean) checkbox data widget.

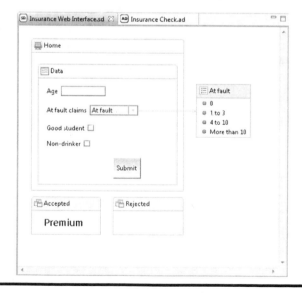

Figure 16.3 Structure diagram for the insurance premium problem.

The automatically generated test cases can be exported into different document and test script formats. The example shown in Table 16.1 is the default output generated by an Excel exporter, that is, test documentation in Excel format. Figure 16.1 shows a list of tools supported at the time of writing. For each test case, the Excel exporter generates information such as test case name and summary, as well as a description of each test step and expected results.

Table 16.1   Generated Test Case (First 2 of 7) Documentation for the Insurance Premium Problem

Test case 1:	Decision/'Insurance Check' | Decision 'Invalid age'/Condition ((Age in 'Entered Data' Variable < 16) or (Age in 'Entered Data' Variable > 90)) is true	
Step	**Action(s)**	**Verification Point(s)**
1	Fill Data form in Home screen entering 0 in Age textbox	No errors can be observed at the SUT
2	Click Submit button in the Home screen in the Data form	Application displays Rejected popup where Text is set to "Sorry, you are not eligible for insurance"
Test case 2:	Decision/'Insurance Check' | Decision 'Between 16 and 25'/ Condition ((Age in 'Entered Data' Variable < 25) and (Age in 'Entered Data' Variable >= 16)) is true	
Step	**Action(s)**	**Verification Point(s)**
1	Fill Data form in Home screen entering 20 in Age textbox selecting "More than 10" from At fault claims dropdown	No errors can be observed at the SUT
2	Click Submit button in the Home screen in the Data form	Application displays Rejected popup where Text is set to "Sorry, you are not eligible for insurance"

16.2.2 Generated Test Cases

Table 16.2 contains a complete, and reformatted, version of test case 1. Conformiq Creator performs a coverage analysis of the activity diagram with respect to four criteria: activity diagram nodes, action invocations, control flows, decision conditions and data. These results are also exported to a separate Traceability Matrix worksheet by the Excel importer (see Tables 16.3 through 16.6). (These tables are reformatted to improve readability.) Note that at the time of writing, Conformiq Creator did

Table 16.2   Complete Version of Test Case 1

Test case 1:	Decision/"Insurance Check' | Decision 'Invalid age'/ Condition ((Age in 'Entered Data' Variable < 16) or (Age in 'Entered Data' Variable > 90)) is true			
Summary:	*Fill in*			
Overall Verdict:	Open	**Executed against SUT Release:**	*Fill in*	
Executed by:	*Fill in*	**Test Execution date and time:**	*Fill in*	
Step	**Action(s)**	**Verification Point(s)**	**Verdict**	**Observations**
1	Fill Data form in Home screen entering 0 in Age textbox	No errors can be observed at the SUT	Open	*Fill in*
2	Click Submit button in the Home screen in the Data form	Application displays Rejected popup where Text is set to "Sorry, you are not eligible for insurance"	Open	*Fill in*

not yet exploit the full coverage analyses that are supported by the Conformiq test generation engine. Other options include boundary value analysis, multiple condition/multiple decision, 2-transitions, and all-paths coverage. Conformiq Creator also produces an extensive set of coverage and traceability information with respect to the model from which the test cases are derived. Tables 16.3 through 16.6 report the test case coverage of

- Action invocations (Table 16.3).
- Control flows (Table 16.4).
- Nodes (in the flowchart) (Table 16.4).
- Decisions (Table 16.3).

The text in the column headers in Tables 16.3 through 16.6 is as follows (the single letter abbreviations are used to permit more narrow columns).

16.2.3 Test Coverage Analysis

The text in the column headers in Tables 16.3 through 16.6 is as follows (the single letter abbreviations are used to permit more narrow columns):

A Decision/'Insurance Check' | Decision 'Invalid age'/Condition ((Age in 'Entered Data' Variable < 16) or (Age in 'Entered Data' Variable > 90)) is true

B Decision/'Insurance Check' | Decision 'Between 16 and 25'/Condition ((Age in 'Entered Data' Variable < 25) and (Age in 'Entered Data' Variable >= 16)) is true

C Decision/'Insurance Check' | Decision 'Older than 65'/Condition (Age in 'Entered Data' Variable >= 65) is false

D Decision/'Insurance Check' | Decision 'goodStudent'/Condition (goodStudent in 'Entered Data' Variable = true) is false

E Decision/'Insurance Check' | Decision 'Non-drinker'/Condition (Non-drinker in 'Entered Data' Variable = true) is false

F Decision/'Insurance Check' | Decision '4 to 10 claims'/Condition (At fault claims in 'Entered Data' Variable = 4 to 10) is false

G Decision/'Insurance Check' | Decision '4 to 10 claims'/Condition (At fault claims in 'Entered Data' Variable = 4 to 10) is true

Table 16.3 **(Reformatted) Traceability Matrix of Action Invocations**

Action Invocations	A	B	C	D	E	F	G
'Insurance Check'/Activity 'Base Rate 1.2'/ Action #1: SV: 'Premium' <- 360				X	X	X	X
'Insurance Check'/Activity 'Base Rate 1.5'/ Action #1: SV: 'Premium' <- 450		X					
'Insurance Check'/Activity 'Get inputs'/Action #1: FF: 'Data' Form	X	X	X	X	X	X	X
'Insurance Check'/Activity 'Get inputs'/Action #2: CB: 'Submit' Button	X	X	X	X	X	X	X
'Insurance Check'/Activity 'No insurance'/ Action #1: DP: 'Rejected' Popup	X	X	X				
'Insurance Check'/Activity 'Normal base rate'/ Action #1: SV: 'Premium' <- 300			X				
'Insurance Check'/Activity 'Premium plus 100'/Action #1: SV: 'Premium' <- ('Premium' Variable + 100)				X	X		
'Insurance Check'/Activity 'Premium plus 300'/Action #1: SV: 'Premium' <- ('Premium' Variable + 300)							X
'Insurance Check'/Activity 'Reduce for goodStudent'/Action #1: SV: 'Premium' <- ('Premium' Variable - 50)					X	X	X
'Insurance Check'/Activity 'Reduce for non-drinker'/Action #1: SV: 'Premium' <- ('Premium' Variable - 75)				X		X	X
'Insurance Check'/Activity 'You have been accepted to our insurance program and your premium will be $'/Action #1: DP: 'Accepted' Popup				X	X	X	X

Table 16.4 (Reformatted) Traceability Matrix of Control Flows

Control Flows	A	B	C	D	E	F	G
'Insurance Check'/Activity 'Base Rate 1.2' -> Decision 'More than 10 claims'				X	X	X	X
'Insurance Check'/Activity 'Base Rate 1.5' -> Decision 'More than 10 claims'		X					
'Insurance Check'/Activity 'Get inputs' -> Decision 'Invalid age'	X	X	X	X	X	X	X
'Insurance Check'/Activity 'Normal base rate' -> Decision 'More than 10 claims'			X				
'Insurance Check'/Activity 'Premium plus 100' -> Decision 'Good student'				X	X		
'Insurance Check'/Activity 'Premium plus 300' -> Decision 'Good student'							X
'Insurance Check'/Activity 'Reduce for good student' -> Decision 'Non-drinker'					X	X	X
'Insurance Check'/Activity 'Reduce for non-drinker' -> Activity 'You have been accepted to our insurance program and your premium will be $'				X		X	X
'Insurance Check'/Decision '1 to 3 claims' -> Activity 'Premium plus 100' (YES)				X	X		
'Insurance Check'/Decision '1 to 3 claims' -> Decision '4 to 10 claims' (NO)						X	X
'Insurance Check'/Decision '4 to 10 claims' -> Activity 'Premium plus 300' (YES)							X
'Insurance Check'/Decision '4 to 10 claims' -> Decision 'Good student' (NO)						X	
'Insurance Check'/Decision 'Between 16 and 25' -> Activity 'Base Rate 1.5' (YES)		X					
'Insurance Check'/Decision 'Between 16 and 25' -> Decision 'Older than 65' (NO)			X	X	X	X	X
'Insurance Check'/Decision 'Good student' -> Activity 'Reduce for good student' (YES)					X	X	X

(*Continued*)

Table 16.4 (*Continued*)   (Reformatted) Traceability Matrix of Control Flows

Control Flows	A	B	C	D	E	F	G
'Insurance Check'/Decision 'Good student' -> Decision 'Non-drinker' (NO)				X			
'Insurance Check'/Decision 'Invalid age' -> Activity 'No insurance' (YES)	X						
'Insurance Check'/Decision 'Invalid age' -> Decision 'Between 16 and 25' (NO)		X	X	X	X	X	X
'Insurance Check'/Decision 'More than 10 claims' -> Activity 'No insurance' (YES)		X	X				
'Insurance Check'/Decision 'More than 10 claims' -> Decision '1 to 3 claims' (NO)				X	X	X	X
'Insurance Check'/Decision 'Non-drinker' -> Activity 'Reduce for non-drinker' (YES)				X		X	X
'Insurance Check'/Decision 'Non-drinker' -> Activity 'You have been accepted to our insurance program and your premium will be $' (NO)					X		
'Insurance Check'/Decision 'Older than 65' -> Activity 'Base Rate 1.2' (YES)				X	X	X	X
'Insurance Check'/Decision 'Older than 65' -> Activity 'Normal base rate' (NO)			X				
'Insurance Check'/Initial -> Activity 'Get inputs'	X	X	X	X	X	X	X

Table 16.5   (Reformatted) Traceability Matrix of Activity Diagram Nodes

Activity Diagram Nodes	A	B	C	D	E	F	G
'Insurance Check'/Activity 'Base Rate 1.2'				X	X	X	X
'Insurance Check'/Activity 'Base Rate 1.5'		X					
'Insurance Check'/Activity 'Get inputs'	X	X	X	X	X	X	X
'Insurance Check'/Activity 'No insurance'	X	X	X				
'Insurance Check'/Activity 'Normal base rate'			X				

(Continued)

Table 16.5 (*Continued*) (Reformatted) Traceability Matrix of Activity Diagram Nodes

Activity Diagram Nodes	A	B	C	D	E	F	G
'Insurance Check'/Activity 'Premium plus 100'				X	X		
'Insurance Check'/Activity 'Premium plus 300'							X
'Insurance Check'/Activity 'Reduce for good student'					X	X	X
'Insurance Check'/Activity 'Reduce for non-drinker'				X		X	X
'Insurance Check'/Activity 'You have been accepted to our insurance program and your premium will be $'				X	X	X	X
'Insurance Check'/Decision '1 to 3 claims'				X	X	X	X
'Insurance Check'/Decision '4 to 10 claims'						X	X
'Insurance Check'/Decision 'Between 16 and 25'		X	X	X	X	X	X
'Insurance Check'/Decision 'Good student'				X	X	X	X
'Insurance Check'/Decision 'Invalid age'	X	X	X	X	X	X	X
'Insurance Check'/Decision 'More than 10 claims'		X	X	X	X	X	X
'Insurance Check'/Decision 'Non-drinker'				X	X	X	X
'Insurance Check'/Decision 'Older than 65'			X	X	X	X	X
'Insurance Check'/Initial	X	X	X	X	X	X	X

Table 16.6 (Reformatted) Traceability Matrix of Decision Coverage

Decision Conditions	A	B	C	D	E	F	G
Decision/'Insurance Check' | Decision '1 to 3 claims'/Condition (At fault claims in 'Entered Data' Variable = 1 to 3) is false						X	X
Decision/'Insurance Check' | Decision '1 to 3 claims'/Condition (At fault claims in 'Entered Data' Variable = 1 to 3) is true				X	X		

(*Continued*)

Table 16.6 (*Continued*)   (Reformatted) Traceability Matrix of Decision Coverage

Decision Conditions	A	B	C	D	E	F	G
Decision/'Insurance Check' | Decision '4 to 10 claims'/Condition (At fault claims in 'Entered Data' Variable = 4 to 10) is false						X	
Decision/'Insurance Check' | Decision '4 to 10 claims'/Condition (At fault claims in 'Entered Data' Variable = 4 to 10) is true							X
Decision/'Insurance Check' | Decision 'Between 16 and 25'/ Condition ((Age in 'Entered Data' Variable < 25) and (Age in 'Entered Data' Variable >= 16)) is false			X	X	X	X	X
Decision/'Insurance Check' | Decision 'Between 16 and 25'/ Condition ((Age in 'Entered Data' Variable < 25) and (Age in 'Entered Data' Variable >= 16)) is true		X					
Decision/'Insurance Check' | Decision 'Good student'/Condition (Good student in 'Entered Data' Variable = true) is false				X			
Decision/'Insurance Check' | Decision 'Good student'/Condition (Good student in 'Entered Data' Variable = true) is true					X	X	X
Decision/'Insurance Check' | Decision 'Invalid age'/Condition ((Age in 'Entered Data' Variable < 16) or (Age in 'Entered Data' Variable > 90)) is false		X	X	X	X	X	X
Decision/'Insurance Check' | Decision 'Invalid age'/Condition ((Age in 'Entered Data' Variable < 16) or (Age in 'Entered Data' Variable > 90)) is true	X						

(*Continued*)

Table 16.6 (*Continued*)   (Reformatted) Traceability Matrix of Decision Coverage

Decision Conditions	A	B	C	D	E	F	G
Decision/'Insurance Check' | Decision 'More than 10 claims'/Condition (At fault claims in 'Entered Data' Variable = More than 10) is false				X	X	X	X
Decision/'Insurance Check' | Decision 'More than 10 claims'/Condition (At fault claims in 'Entered Data' Variable = More than 10) is true		X	X				
Decision/'Insurance Check' | Decision 'Non-drinker'/Condition (Non-drinker in 'Entered Data' Variable = true) is false					X		
Decision/'Insurance Check' | Decision 'Non-drinker'/Condition (Non-drinker in 'Entered Data' Variable = true) is true				X		X	X
Decision/'Insurance Check' | Decision 'Older than 65'/Condition (Age in 'Entered Data' Variable >= 65) is false			X				
Decision/'Insurance Check' | Decision 'Older than 65'/Condition (Age in 'Entered Data' Variable >= 65) is true				X	X	X	X

The complete reports of Traceability Matrix information are summarized in a test suite report (copied in Table 16.7).

Conformiq provided test cases for different alternative configurations of test generation. Results are summarized in Tables 16.7 and 16.8. The two sets of test cases show some of the flexibility in test case generation and illustrate the difference in generating the most compact test suite with least tests and test steps versus a test suite complying with a well-established conformance testing standard ISO-9646 [ISO 1991]. Table 16.9 lists a second set of ten test cases that are compatible with the well-established conformance testing standard ISO-9646. Table 16.10 contains a detailed view of one of the test cases in Table 16.9. In the third test set provided, all combinations of input data were selected generating 64 tests.

Table 16.7 Test Suite Summary

Conformiq Test Specification	
Project	Insurance Premium Calculation
Generation Date	*7.3.2016 8:26*
Number of Tests Exported	7
Conformiq Options	
Automatic test case naming	True
Lookahead depth	1
OSI methodology support	False
Only finalized runs	Disabled
Coverage Summary	
Activity diagrams	100% (55/55)
Decision	100% (16/16)
Overall coverage	100% (71/71)

Table 16.8 Condensed Version of the Seven Test Cases in Table 16.1

Test Case	Age	Claims	goodStudent	nonDrinker	Result
1	0	—	—	—	Not eligible for insurance
2	20	>10	—	—	Not eligible for insurance
3	44	>10	—	—	Not eligible for insurance
4	77	1–3	False	True	Premium is $385
5	77	1–3	True	False	Premium is $410
6	77	0	True	True	Premium is $235
7	77	4–10	True	False	Premium is $535

Tables 16.11 and 16.12 contain the 64 (reformatted) test cases from all the combinations of input data option. The test case order has been reorganized into a more discernable way in Table 16.12. These 64 test cases constitute Worst Case Normal equivalence class testing [Jorgensen 2014].

Table 16.9   Condensed Version of the Submitted Set of 10 Test Cases

Test Case	Age	Claims	goodStudent	nonDrinker	Result
1	0	—	—	—	Not eligible for insurance
2	0	—	—	—	Not eligible for insurance
3	20	>10	—	—	Not eligible for insurance
4	77	>10	—	—	Not eligible for insurance
5	44	>10	—	—	Not eligible for insurance
6	77	1–3	True	False	Premium is $410
7	77	1–3	True	True	Premium is $335
8	77	1–3	False	True	Premium is $385
9	77	0	True	True	Premium is $235
10	77	4–10	True	True	Premium is $535

Table 16.10   Sample Test Case from Generated Test Documentation for the Insurance Premium Problem

Test Case 4:	Decision/"Insurance Check" | Decision 'Good student'/Condition (goodStudent in 'Entered Data')	
Summary:	*Fill in*	
Overall Verdict:	Open	**Executed against SUT Release:**
Executed by:	*Fill in*	**Test Execution date and time:**
Step	**Action(s)**	**Verification Point(s)**
1	Fill Data form in Home screen entering 77 in Age textbox selecting "1 to 3" from At fault claims dropdown unchecking goodStudent checkbox checking nonDrinker checkbox	No errors can be observed at the SUT

(Continued)

Table 16.10 (*Continued*)   Sample Test Case from Generated Test Documentation for the Insurance Premium Problem

2	Click Submit button in the Home screen in the Data form	Application displays Accepted popup where Text is set to "Client can be ensured" Premium label is set to "Your premium will be $385"

Table 16.11   Sixty-Four Generated Insurance Problem Test Cases

Test Case	Age	Claims	goodStudent	nonDrinker	Approved?	Premium
1	77	1 to 3	F	T		$385
2	77	1 to 3	T	F		$410
3	77	0	T	T		$235
4	77	4 to 10	T	T		$535
5	0	0	F	F	NO	
6	0	1 to 3	F	F	NO	
7	0	4 to 10	F	F	NO	
8	0	4 to 10	F	T	NO	
9	0	4 to 10	T	F	NO	
10	0	>10	F	F	NO	
11	0	>10	F	T	NO	
12	0	>10	T	F	NO	
13	77	>10	T	T	NO	
14	20	0	T	T		$325
15	77	0	F	T		$285
16	77	0	T	F		$310
17	44	0	T	T		$175
18	77	1 to 3	T	T		$335

(Continued)

Table 16.11 (*Continued*) Sixty-Four Generated Insurance Problem Test Cases

Test Case	Age	Claims	goodStudent	nonDrinker	Approved?	Premium
19	0	0	F	T	NO	
20	0	0	T	F	NO	
21	0	0	T	T	NO	
22	0	1 to 3	F	T	NO	
23	0	1 to 3	T	F	NO	
24	0	1 to 3	T	T	NO	
25	0	4 to 10	T	T	NO	
26	0	>10	T	T	NO	
27	20	>10	F	F	NO	
28	20	>10	F	T	NO	
29	20	>10	T	F	NO	
30	20	>10	T	T	NO	
31	44	>10	F	F	NO	
32	44	>10	F	T	NO	
33	44	>10	T	F	NO	
34	44	>10	T	T	NO	
35	77	>10	F	F	NO	
36	77	>10	F	T	NO	
37	77	>10	T	T	NO	
38	20	0	F	F		$450
39	20	1 to 3	F	F		$550
40	20	0	F	T		$375
41	20	0	T	F		$400

(*Continued*)

Table 16.11 (*Continued*)   Sixty-Four Generated Insurance Problem Test Cases

Test Case	Age	Claims	goodStudent	nonDrinker	Approved?	Premium
42	20	1 to 3	F	T		$475
43	20	1 to 3	T	F		$500
44	20	4 to 10	F	F		$750
45	44	0	F	F		$300
46	44	1 to 3	F	F		$400
47	77	0	F	F		$360
48	77	1 to 3	F	F		$460
49	20	1 to 3	T	T		$425
50	20	4 to 10	F	T		$675
51	20	4 to 10	T	F		$700
52	44	0	F	T		$225
53	44	0	T	F		$250
54	44	1 to 3	F	T		$325
55	44	1 to 3	T	F		$350
56	44	4 to 10	T	F		$600
57	77	4 to 10	F	F		$660
58	20	4 to 10	T	T		$625
59	44	1 to 3	T	T		$275
60	44	4 to 10	F	T		$525
61	44	4 to 10	T	F		$550
62	77	4 to 10	F	T		$585
63	77	4 to 10	T	F		$610
64	44	4 to 10	T	T		$475

Table 16.12 Sixty-Four (Reorganized) Generated Insurance Problem Test Cases

Test Case	Age	Claims	goodStudent	nonDrinker	Approved?	Premium
5	0	0	F	F	NO	
19	0	0	F	T	NO	
20	0	0	T	F	NO	
21	0	0	T	T	NO	
10	0	>10	F	F	NO	
11	0	>10	F	T	NO	
12	0	>10	T	F	NO	
26	0	>10	T	T	NO	
6	0	1 to 3	F	F	NO	
22	0	1 to 3	F	T	NO	
23	0	1 to 3	T	F	NO	
24	0	1 to 3	T	T	NO	
7	0	4 to 10	F	F	NO	
8	0	4 to 10	F	T	NO	
9	0	4 to 10	T	F	NO	
25	0	4 to 10	T	T	NO	
27	20	>10	F	F	NO	
28	20	>10	F	T	NO	
29	20	>10	T	F	NO	
30	20	>10	T	T	NO	
31	44	>10	F	F	NO	
32	44	>10	F	T	NO	
33	44	>10	T	F	NO	

(*Continued*)

Table 16.12 (*Continued*) Sixty-Four (Reorganized) Generated Insurance Problem Test Cases

Test Case	Age	Claims	goodStudent	nonDrinker	Approved?	Premium
34	44	>10	T	T	NO	
13	77	>10	T	T	NO	
35	77	>10	F	F	NO	
36	77	>10	F	T	NO	
37	77	>10	T	T	NO	
14	20	0	T	T		$325
38	20	0	F	F		$450
40	20	0	F	T		$375
41	20	0	T	F		$400
39	20	1 to 3	F	F		$550
42	20	1 to 3	F	T		$475
43	20	1 to 3	T	F		$500
49	20	1 to 3	T	T		$425
44	20	4 to 10	F	F		$750
50	20	4 to 10	F	T		$675
51	20	4 to 10	T	F		$700
58	20	4 to 10	T	T		$625
17	44	0	T	T		$175
45	44	0	F	F		$300
52	44	0	F	T		$225
53	44	0	T	F		$250
46	44	1 to 3	F	F		$400
54	44	1 to 3	F	T		$325

(*Continued*)

Table 16.12 (*Continued*) Sixty-Four (Reorganized) Generated Insurance Problem Test Cases

Test Case	Age	Claims	goodStudent	nonDrinker	Approved?	Premium
55	44	1 to 3	T	F		$350
59	44	1 to 3	T	T		$275
56	44	4 to 10	F	F		$600
60	44	4 to 10	F	T		$525
61	44	4 to 10	T	F		$550
64	44	4 to 10	T	T		$475
3	77	0	T	T		$235
15	77	0	F	T		$285
16	77	0	T	F		$310
47	77	0	F	F		$360
1	77	1 to 3	F	T		$385
2	77	1 to 3	T	F		$410
18	77	1 to 3	T	T		$335
48	77	1 to 3	F	F		$460
4	77	4 to 10	T	T		$535
57	77	4 to 10	F	F		$660
62	77	4 to 10	F	T		$585
63	77	4 to 10	T	F		$610

16.3 Garage Door Controller Results

The Conformiq Designer tool was used on the Garage Door Controller problem. The first two steps were to create a UML state machine diagram (Figure 16.4) and to complement CQA files (Conformiq's UML action language which is based on Java) to define the interfaces available for testing (system.cqa), methods called from the statechart (GarageDoorController.cqa), as well as the instantiation of the model component (main.cqa) (Figure 16.4).

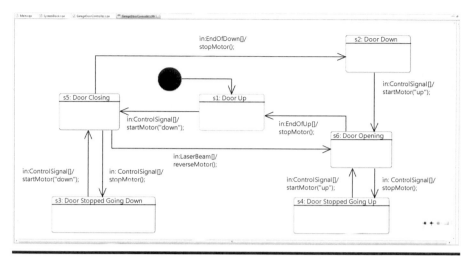

Figure 16.4 **Garage Door Controller state machine diagram.**

16.3.1 Input Diagram and QML Text Files

The Conformiq state machine diagram in Figure 16.4 is an extended finite state machine; it is topologically equivalent to the one in Chapter 13. Compared to Conformiq Creator, Conformiq Designer uses a programmatic notation for defining interfaces and expressing actions and conditions. As in the insurance premium case, Conformiq Designer automatically derives test cases from the state machine diagram. In one run of the program, 17 test cases were generated, which are shown in Tables 16.13 and 16.14. Despite the difference in modeling notation, the generated tests can also be exported in the same Excel-based test documentation format. As earlier, we show only the most relevant subset of data.

16.3.1.1 CQA Files

The following lists the files that complement and define methods referred from the state machine diagram shown in Figure 16.4.

```
/** Main.cqa: entry point to the modeled behavior – here a single model
component. */
void main()
{
    // Instantiate and start execution of 'GarageDoorController' model component.
    GarageDoorController mc = new GarageDoorController();
    mc.start("GarageDoorController");
}
```

```
/** System.cqa: Declaration of the external interface of the system being modeled.
*/
system
{
    Inbound in : ControlSignal, EndOfDown, EndOfUp, LaserBeam;
    Outbound out : MotorStart, MotorStop, MotorReverse;
}

record ControlSignal { }
record EndOfDown { }
record EndOfUp { }
record LaserBeam { }
record MotorStart { String direction; }
record MotorStop { }
record MotorReverse { }

/** GarageDoorController.cqa: Declaration of 'GarageDoorController' state
machine instance. */
class GarageDoorController extends StateMachine
{
    /** The default constructor. */
    public GarageDoorController() { }

    public void startMotor( String direction ) {
        MotorStart start;
        start.direction = direction;
        out.send( start );
    }

    public void stopMotor( ) {
        MotorStop stop;
        out.send( stop );
    }

    public void reverseMotor( ) {
        MotorReverse reverse;
        out.send( reverse );
    }
}
```

The state sequences in these 17 test cases use the following abbreviations (from Chapter 13): s1: Door Up, s2: Door Down, s3: Door stopped going down, s4: Door stopped going up, s5: Door closing, and s6: Door opening.

16.3.2 Generated Test Cases

Table 16.13 shows the 17 test cases generated for the Garage Door Controller. (The original format of just one test case is shown in Table 16.14.) The expanded test cases in Table 16.13 show the state sequences covered by each test case.

Table 16.13 Seventeen Test Cases Generated from the Garage Door Controller State Machine Diagram

Step	Action(s)	Verification Point(s)
Test case 1:	s1, s5	
1	Provide system with input e1: control signal	System performs action a1: start drive motor "down".
Test case 2:	s1, s5, s3	
1	Provide system with input e1: control signal	System performs action a1: start drive motor "down".
2	Provide system with input e1: control signal	System performs action a3: stop drive motor.
Test case 3:	s1, s5, s2	
1	Provide system with input e1: control signal	System performs action a: start drive motor "down".
2	Provide system with input e2: end of down track hit	System performs action a3: stop drive motor.
Test case 4:	s1, s5, s3, s5	
1	Provide system with input e1: control signal	System performs action a1: start drive motor "down".
2	Provide system with input e1: control signal	System performs action a3: stop drive motor.
3	Provide system with input e1: control signal	System performs action a1: start drive motor "down".
Test case 5:	s1, s5, s6	
1	Provide system with input e1: control signal	System performs action a1: start drive motor "down".
2	Provide system with input e4: laser beam crossed	System performs action a4: reverse motor down to up.

(Continued)

Table 16.13 (*Continued*)   Seventeen Test Cases Generated from the Garage Door Controller State Machine Diagram

Step	Action(s)	Verification Point(s)
Test case 6:	s1, s5, s2, s6	
1	Provide system with input e1: control signal	System performs action a1: start drive motor "down".
2	Provide system with input e2: end of down track hit	System performs action a3: stop drive motor.
3	Provide system with input e1: control signal	System performs action a2: start drive motor "up".
Test case 7:	s1, s5, s6, s1	
1	Provide system with input e1: control signal	System performs action a1: start drive motor "down".
2	Provide system with input e4: laser beam crossed	System performs action a4: reverse motor down to up.
3	Provide system with input e3: end of up track hit	System performs action a3: stop drive motor.
Test case 8:	s1,s5, s3, s5, s3	
1	Provide system with input e1: control signal	System performs action a1: start drive motor "down".
2	Provide system with input e1: control signal	System performs action a3: stop drive motor.
3	Provide system with input e1: control signal	System performs action a1: start drive motor "down".
4	Provide system with input e1: control signal	System performs action a3: stop drive motor.
Test case 9:	s1, s5, s3, s5, s2	
1	Provide system with input e1: control signal	System performs action a1: start drive motor "down".
2	Provide system with input e1: control signal	System performs action a3: stop drive motor.
3	Provide system with input e1: control signal	System performs action a1: start drive motor "down".

(*Continued*)

Table 16.13 (*Continued*) Seventeen Test Cases Generated from the Garage Door Controller State Machine Diagram

Step	Action(s)	Verification Point(s)
4	Provide system with input e2: end of down track hit	System performs action a3: stop drive motor.
Test case 10:	s1, s5, s3, s5, s6	
1	Provide system with input e1: control signal	System performs action a1: start drive motor "down".
2	Provide system with input e1: control signal	System performs action a3: stop drive motor.
3	Provide system with input e1: control signal	System performs action a1: start drive motor "down".
4	Provide system with input e4: laser beam crossed	System performs action a4: reverse motor down to up.
Test case 11:	s1, s5, s2, s6, s1	
1	Provide system with input e1: control signal	System performs action a1: start drive motor "down".
2	Provide system with input e2: end of down track hit	System performs action a3: stop drive motor.
3	Provide system with input e1: control signal	System performs action a2: start drive motor "up".
4	Provide system with input e3: end of up track hit	System performs action a3: stop drive motor.
Test case 12:	s1, s5, s6, s1, s5	
1	Provide system with input e1: control signal	System performs action a1: start drive motor "down".
2	Provide system with input e4: laser beam crossed	System performs action a4: reverse motor down to up.
3	Provide system with input e3: end of up track hit	System performs action a3: stop drive motor.
4	Provide system with input e1: control signal	System performs action a1: start drive motor "down".

(Continued)

Table 16.13 (*Continued*) Seventeen Test Cases Generated from the Garage Door Controller State Machine Diagram

Step	Action(s)	Verification Point(s)
Test case 13:	s1, s5, s6, s4	
1	Provide system with input e1: control signal	System performs action a1: start drive motor "down".
2	Provide system with input e4: laser beam crossed	System performs action a4: reverse motor down to up.
3	Provide system with input e1: control signal	System performs action a3: stop drive motor.
Test case 14:	s1, s5, s2, s6, s4	
1	Provide system with input e1: control signal	System performs action a1: start drive motor "down".
2	Provide system with input e2: end of down track hit	System performs action a3: stop drive motor.
3	Provide system with input e1: control signal	System performs action a2: start drive motor "up".
4	Provide system with input e1: control signal	System performs action a3: stop drive motor.
Test case 15:	s1, s5, s6, s4, s6	
1	Provide system with input e1: control signal	System performs action a1: start drive motor "down".
2	Provide system with input e4: laser beam crossed.	System performs action a4: reverse motor down to up.
3	Provide system with input e1: control signal	System performs action a3: stop drive motor.
4	Provide system with input e1: control signal	System performs action a2: start drive motor "up".
Test case 16:	s1, s5, s6, s4, s6, s4	
1	Provide system with input e1: control signal	System performs action a1: start drive motor "down".
2	Provide system with input e4: laser beam crossed	System performs action a4: reverse motor down to up.

(Continued)

Table 16.13 (*Continued*)   Seventeen Test Cases Generated from the Garage Door Controller State Machine Diagram

Step	Action(s)	Verification Point(s)
3	Provide system with input e1: control signal	System performs action a3: stop drive motor.
4	Provide system with input e1: control signal	System performs action a2: start drive motor "up".
5	Provide system with input e1: control signal	
Test case 17:	s1, s5, s6, s4, s6, s1	
1	Provide system with input e1: control signal	System performs action a1: start drive motor "down".
2	Provide system with input e4: laser beam crossed	System performs action a4: reverse motor down to up.
3	Provide system with input e1: control signal	System performs action a3: stop drive motor.
4	Provide system with input e1: control signal	System performs action a2: start drive motor "up".
5	Provide system with input e3: end of up track hit	System performs action a3: stop drive motor.

Table 16.14   Sample Test Case Provided by Conformiq in Original Format

Test case 5:	Door up, Door Closing, Door Opening	
Step	**Action(s)**	**Verification Point(s)**
1	Provide system with input ControlSignal via in.	System responds with MotorStart via out where direction is "down".
2	Provide system with input LaserBeam via in.	System responds with MotorReverse via out.

16.3.3 Traceability Matrices

The data provided in Tables 16.16 through 16.20 shows the coverage of state machine diagram information such as states, transitions, 2-transitions, methods, and statements. Since the example does not contain any logic on data, no coverage for decision, compound condition, MC/DC, or boundary value coverage could be generated. The test configuration in the example was configured to produce tests compliant to ISO 9646 conformance testing methodology framework. The 17 test cases are shown in Table 16.15 for reference in Tables 16.16 through 16.20.

Table 16.15   Test Case Numbers

Test Case	State Sequence	Test Case	State Sequence
1	s1, s5	10	s1, s5, s3, s5, s6
2	s1, s5, s3	11	s1, s5, s2, s6, s1
3	s1, s5, s2	12	s1, s5, s6, s1, s5
4	s1, s5, s3, s5	13	s1, s5, s6, s4
5	s1, s5, s6	14	s1, s5, s2, s6, s4
6	s1, s5, s2, s6	15	s1, s5, s6, s4, s6
7	s1, s5, s6, s1	16	s1, s5, s6, s4, s6, s4
8	s1, s5, s3, s5, s3	17	s1, s5, s6, s4, s6, s1
9	s1, s5, s3, s5, s2		

Table 16.16   Traceability Matrix of States

Test Case Number	1	2	3	4	5	6	7	8	9	10	11	12	13	14	15	16	17
States																	
Initial state	X	X	X	X	X	X	X	X	X	X	X	X	X	X	X	X	X
s1: Door Up	X	X	X	X	X	X	X	X	X	X	X	X	X	X	X	X	X
s2: Door Down			X		X			X		X				X			
s3: Door Stopped Going Down		X		X				X	X	X							
s4: Door Stopped Going Up													X	X	X	X	X
s5: Door Closing	X	X	X	X	X	X	X	X	X	X	X	X	X	X	X	X	X
s6: Door Opening					X	X	X			X	X	X	X	X	X	X	X

Table 16.17 Traceability Matrix of Transitions

Test Case Number	1	2	3	4	5	6	7	8	9	10	11	12	13	14	15	16	17
Transitions																	
Initial state -> s1: Door Up [0]	X	X	X	X	X	X	X	X	X	X	X	X	X	X	X	X	X
s1: Door Up -> s5: Door Closing [1]	X	X	X	X	X	X	X	X	X	X	X	X	X	X	X	X	X
s2: Door Down -> s6: Door Opening [5]						X								X			
s3: Door Stopped Going Down -> s5: Door Closing [3]				X							X						
s4: Door Stopped Going Up -> s6: Door Opening [8]										X					X	X	X
s5: Door Closing -> s2: Door Down [4]			X			X			X		X			X			
s5: Door Closing -> s3: Door Stopped Going Down [2]		X		X				X	X	X							
s5: Door Closing -> s6: Door Opening [6]					X		X			X		X	X		X	X	X
s6: Door Opening -> s1: Door Up [9]							X				X	X	X			X	X
s6: Door Opening -> s4: Door Stopped Going Up [7]													X	X	X	X	X

Table 16.18 Traceability Matrix of 2-Transitions (First 5 Transition Pairs Only)

Test Case Number	1	2	3	4	5	6	7	8	9	10	11	12	13	14	15	16	17
2-Transitions																	
initial state -> s1: Door Up -> s5: Door Closing [0:1]	X	X	X	X	X	X	X	X	X	X	X	X	X	X	X	X	X
s1: Door Up -> s5: Door Closing -> s2: Door Down [1:4]			X			X					X			X			
s1: Door Up -> s5: Door Closing -> s3: Door Stopped Going Down [1:2]		X		X				X	X	X							
s1: Door Up -> s5: Door Closing -> s6: Door Opening [1:6]					X		X					X	X		X	X	X
s2: Door Down -> s6: Door Opening -> s1: Door Up [5:9]											X						

Table 16.19 Traceability Matrix of Methods

Test Case Number	1	2	3	4	5	6	7	8	9	10	11	12	13	14	15	16	17
Methods																	
GarageDoorController()	X	X	X	X	X	X	X	X	X	X	X	X	X	X	X	X	X
reverseMotor()					X		X			X		X	X		X	X	X
startMotor(String)	X	X	X	X	X	X	X	X	X	X	X	X	X	X	X	X	X
stopMotor()		X	X	X		X	X	X	X	X	X	X	X	X	X	X	X
Global Scope/main()	X	X	X	X	X	X	X	X	X	X	X	X	X	X	X	X	X

Table 16.20 Traceability Matrix of Statements (First 9 Statements Only)

Test Case Number	1	2	3	4	5	6	7	8	9	10	11	12	13	14	15	16	17
Statements																	
In QML/model/GarageDoorController.cqa:12 statement 'MotorStart start;' [5]	X	X	X	X		X	X	X	X	X	X	X	X	X	X	X	X
In QML/model/GarageDoorController.cqa:13 statement 'start. direction = direction;' [6]	X	X	X	X		X	X	X	X	X	X	X	X	X	X	X	X
In QML/model/GarageDoorController.cqa:14 statement 'out. send (start);' [7]	X	X	X	X		X	X	X	X	X	X	X	X	X	X	X	X
In QML/model/GarageDoorController.cqa:18 statement 'MotorStop stop;' [8]		X	X	X		X	X	X	X	X	X	X	X	X	X	X	X
In QML/model/GarageDoorController.cqa:19 statement 'out. send (stop);' [9]		X	X	X		X	X	X	X	X	X	X	X	X	X	X	X
In QML/model/GarageDoorController.cqa:23 statement 'MotorReverse reverse;' [3]					X		X			X		X	X		X	X	X
In QML/model/GarageDoorController.cqa:24 statement 'out. send (reverse);' [4]					X		X			X		X	X		X	X	X
In QML/model/GarageDoorController.cqa:9 statement 'super ();' [2]	X	X	X	X		X	X	X	X	X	X	X	X	X	X	X	X
In QML/model/Main.cqa:6 statement 'GarageDoorController mc = new GarageDoorController ();' [0]	X	X	X	X		X	X	X	X	X	X	X	X	X	X	X	X

16.4 Vendor Advice

Model-based testing is a disruptive approach to testing and needs to be managed as such. It requires us to change the way we work and enables a total testing transformation. One of the most challenging steps in adopting MBT—as with any other disruptive technology—is to let go of well-established practices. One example is changing our thinking to go from tests to modeling system operation. Another example is changing test process metrics from measuring the number of tests toward using actual test coverage. Another challenge with known test coverage is that we now not only know what we cover but also what we don't cover. This knowledge changes the way we work and think.

With all the excitement about test generation, it is important to remember that MBT is still about black-box functional testing and that MBT does not replace the need for testers—on the contrary, it gives testers a new weapon to explore the space of infinite possibilities to test a system: Think of it as providing a flamethrower allowing to squash bugs flying around in a dark black box. Nevertheless, many facts also remain: Testing is still an infinite problem—even for a quite simple piece of functionality, for example, the Garage Door Controller; the number of possible tests is infinite, for example, by iterating (forever) through loops. The trick is how to test a system with a minimum set of tests and test steps—this is where good tools will help. In addition, tests are only as good as the model from which they are generated—if you have not modeled a part of functionality, then you will not get tests exploiting it. Therefore, the person who is modeling for testing needs to really understand the operation and quirks of the system or application functionality to be tested. This is why testers are still so very important. Finally, every generated test still needs to be validated—the fact that it is automatically generated makes a test correct (with respect to the model) but not necessarily valid (with respect to the real system)—unless of course we take the view that the model acts as *the* reference implementation. This is why it is so important that MBT tools need to provide features to easily review automatically generated tests, for example, using the Traceability Matrix presented in previous sections. One benefit MBT here offers that we can now review models and generated tests with all stakeholders—long before we have even any implementation prototype—as we can now easily automatically transform models and generated tests to formats that these stakeholders understand best.

When it comes to successful industrial deployment of MBT, it needs to be understood that test generation in itself is just one piece in the puzzle that we call "test automation", and there are many others. Over the past years, MBT (or rather automated test design) has evolved from a pure test generation technology to serve as the central integration piece in automating the entire software development life cycle, that is, the place where all information flows from requirement and test management, and test execution automation can be linked together.

References

[ISO 1991]
International Standards Organization (ISO), *Information Technology, Open Systems Interconnection, Conformance Testing Methodology and Framework*. International Standard IS-9646. ISO, Geneve, 1991.
[Jorgensen 2014]
Jorgensen, Paul, *Software Testing: A Craftsman's Approach, fourth edition*. CRC Press, Taylor & Francis Group, Boca Raton, Florida, 2014.

Chapter 17

Elvior

17.1 Introduction

Most of the material in this chapter was provided either by staff at Elvior or taken (with permission) from the company website. Elvior has its headquarters in Tallinn, Estonia, and has partnerships with VerifySoft in Germany, Easy Global Market in France, Conformiq in Finland (and the United States), and Yashaswini Design Solutions and TCloud Information Technologies in India.

The product line includes the TestCast family:

- TestCast T3—full-featured TTCN-3 development and execution platform.
- TestCast MBT—model-based testing (MBT) tool.
- The company website [http://Elvior.com/] lists the mainline products and services in Sections 17.1.1 and 17.1.2.

17.1.1 Elvior TestCast Tools Family

Elvior is a full-spectrum software development service provider; one of their specialties is TTCN-3 testing services; in this chapter we are more interested in their model-based testing services (based on TestCast MBT). TTCN-3 is the third version of the testing and test control notation. It is an international standard; versions of TTCN have been in use (primarily in Europe) since 1992.

The TestCast MBT product can run on Windows, Macs, or Linux platforms, and uses the Eclipse environment. There are no limitations on the programming languages used in the system under test (SUT), see Figure 17.1.

	Runtime	Professional	MBT
Test suite viewer	✓	✓	✓
Run pre-compiled TTCN-3	✓	✓	✓
MSC logs	✓	✓	✓
TTCN-3 editor		✓	✓
TTCN-3 compiler		✓	✓
TTCN-3 debugger		✓	✓
TTCN-3 tests generation			✓
Test run analysis on model			✓

Figure 17.1 TestCast editions.

TestCast MBT is a bundle composed of

■ The TestCast MBT Client—an Eclipse based frontend for modeling that includes:
  – A State machine editor supporting external data definitions (e.g., TTCN-3).
  – The Test Objective editor for creating structural test coverage criteria.
  – A TestCase(s) viewer.
  – The Test suite implementation for execution platforms (e.g., TTCN-3).
■ The TestCast MBT Server (accessed remotely via the Internet) that
  – Runs on Elvior server and serves multiple TestCast MBT clients.
  – Designs and automatically generates tests according to test coverage criteria provided by TestCast MBT Client.
■ TestCast T3 for automated test execution of TTCN-3 scripts. It is a fully featured TTCN-3 tool for creating test data and automating test execution.

To use the TestCast MBT tool, the user first models the SUT using UML Statecharts, TTCN-3 data definitions, and test configuration information. (The Statecharts are created using the built-in System Model Editor (one example is in Figure 17.3). Next, the desired test coverage is defined, and then TestCast MBT automatically generates abstract test cases. Abstract test sequences can be inspected visually and then rendered as TTCN-3 test scripts, which are executed on the TTCN-3 tool. The test execution results are then available for further analysis.

17.1.2 Testing Related Services

TTCN-3 testing services

  ■ Setup of test environments
  ■ Developing system adapters and codecs

- Developing test suites (TTCN-3 test scripts)
- TTCN-3 training on-site or off-site
- TTCN-2 test suites migration to TTCN-3

Model-based testing services

- Setup test environments
- Model-based testing
- Model-based testing training courses

17.2 Insurance Premium Results

The TestCast MBT tool was used to model system under test (SUT) behavior and generate test cases.

17.2.1 System (SUT) Modeling

A hierarchical model is used for SUT behavior modeling. An EFSM is used for state diagrams (Figure 17.3), whereas data definitions are defined externally as TTCN-3 templates (Figure 17.2).

17.2.2 Test Coverage and Test Generation

The term "test objective" is used in TestCast MBT to define test coverage. Test objectives refer to structural test coverage on the model (e.g., list or set of transitions to visit); there exist predefined test coverage criteria like "all transitions," "all transition pairs," "all transition triplets," and so on. For the insurance premium calculation, a

```
External data – TTCN-3 templates
//TTCN-3 structure for calculating insurance Premium
type record Rule {
        integer age,
        integer atFaultClaims,
        boolean goodStudent,
        boolean nonDrinker
    }
//TTCN-3 template for single rule
//all together there are 39 rule templates to cover Premium calculation logic
//these template are used on SUT model
const Rule rule1 := {
        age := 16,
        atFaultClaims := 0,
        goodStudent := true,
        nonDrinker := true
    }
```

Figure 17.2   TTCN-3 templates for the Insurance Premium Problem.

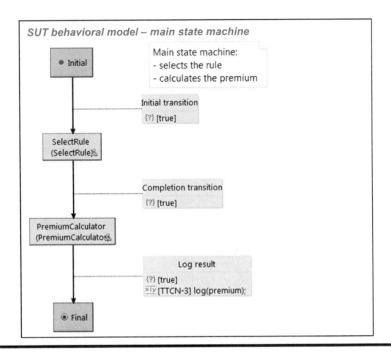

Figure 17.3   High-level workflow for the Insurance Premium Program.

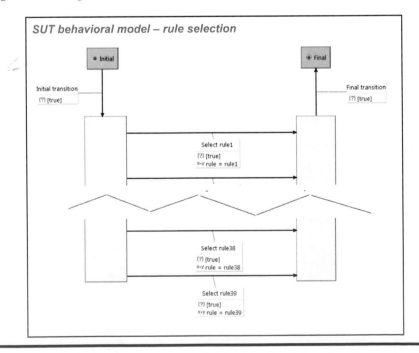

Figure 17.4   Details of rule selection for the Insurance Premium Program.

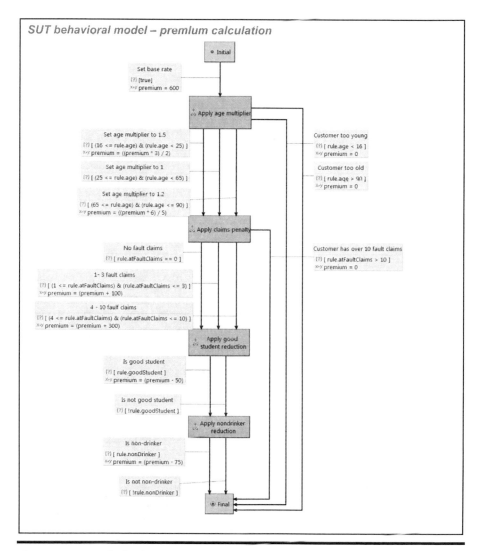

Figure 17.5   Paths in the premium calculation.

test objective covering all 39 rules was created (Figures 17.4 and 17.5). A test objective is set in a way that test generator will first go through a transition for selecting rule and then through SUT final transition "Log result." This is achieved by creating a list of transitions that generator must follow in a specified order. The rest of transition path in between those two transitions is computed automatically according to the "PremiumCalculation" submachine. It is possible to make a list of such transition lists and generating tests for the test objective results in 39 abstract test cases that could be rendered into executable TTCN-3 test scripts (Figures 17.6 through 17.8).

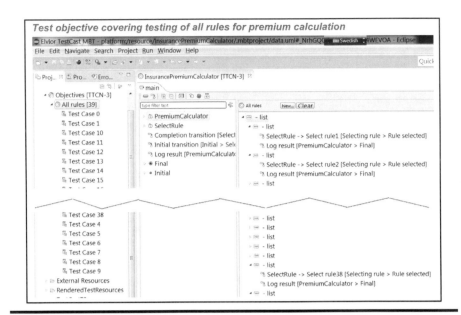

Figure 17.6   Test cases for the *all rules* test objective.

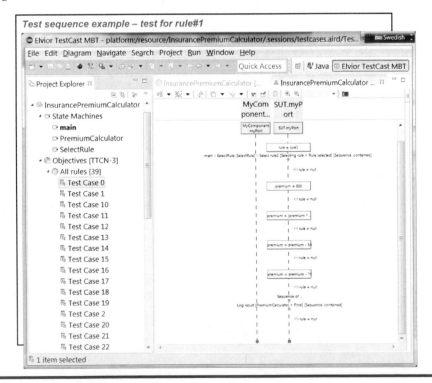

Figure 17.7   Sample test sequence.

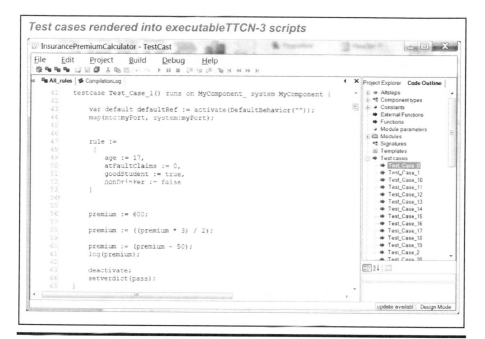

Figure 17.8   Test in TTCN-3 format.

17.3 Garage Door Controller Results

The TestCast MBT tool was used to model system under test (SUT) behavior and generate test cases.

17.3.1 System (SUT) Modeling

A flat model is used for SUT behavior modeling. An EFSM (extended finite state machine) is used for state diagrams, whereas data definitions are defined externally as TTCN-3 templates (Figure 17.9).

17.3.2 Test Coverage and Test Generation

We use the default coverage criteria "all transitions"—this ensures that each transition will be visited at least once. This results in one test case. It is a good idea to use all transitions, because as a test builder, we might not care in which order we test the garage door (go through transitions in model). We just want to know that all functionality works in an expected way. By defining test objective "all transitions," we let the test generator to choose the transition path and we can expect that all functionality is tested.

Since the transition path is selected randomly, sometimes interesting behavior could be discovered for the SUT. As an example, suppose the Garage Door

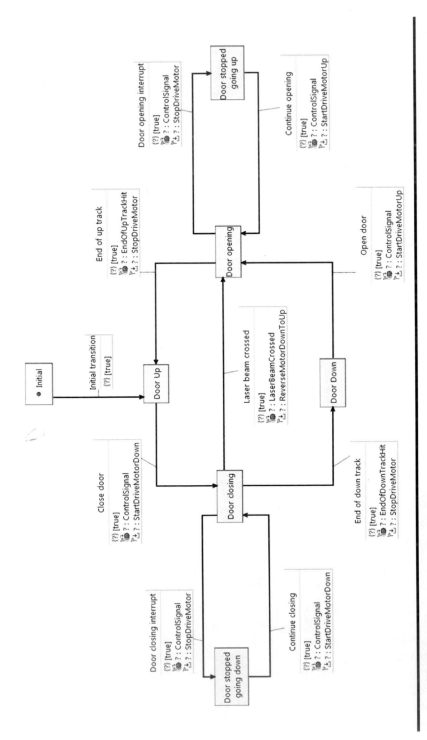

Figure 17.9 Garage Door Controller EFSM.

Controller has a fault such that, when the door is moving down, the user stops it with a control device signal. Next, suppose the user provides another control signal, and the door starts moving down again, but now, the laser beam safety device no longer works, and the door continues closing until it reaches the end of the down track. When writing tests for the Garage Door Controller manually we could make a following test:

1. Close door
2. Laser beam crossed
3. End of up track
4. Close door
5. Door closing interrupt
6. Continue closing
7. End of down track
8. And the rest of transitions related to opening the door

The fault would remain uncaught. As the test generator selects the transition path randomly, then generating all transitions *might* generate a test that first goes through "door closing interrupt," "continue closing," and then "laser beam crossed." The likelihood of discovering the fault can be increased significantly by generating test for "all transition pairs" or even "transition triplets." All pairs would mean that the generated test will have such sequences of transitions (Figures 17.10 through 17.13):

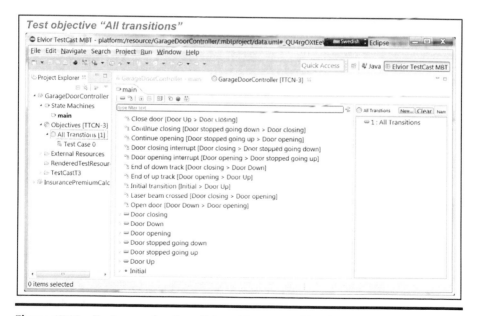

Figure 17.10 Test cases for the *all transitions* test objective.

External data – TTCN-3 types used for modeling of garage door controller
```
type record ControlSignal{}
type record EndOfDownTrackHit{}
type record EndOfUpTrackHit{}
type record LaserBeamCrossed{}
type record StartDriveMotorDown{}
type record StartDriveMotorUp{}
type record StopDriveMotor{}
type record ReverseMotorDownToUp{}
```

Figure 17.11   TTCN-3 data types.

1. Close door → Laser beam crossed
2. Close door → Door closing interrupt
3. Continue closing → End of down track
4. Continue closing → Laser beam crossed
5. The rest of the possible transition pairs

17.4 Vendor Advice

Using the TestCast MBT product requires following skills:

- Basic knowledge of Eclipse—TestCast MBT frontend is Eclipse plugin
- Basic knowledge about TTCN-3—external data definitions used on model are in TTCN-3
- Basic knowledge of UML—SUT behavior is modeled as UML state charts

The tool includes comprehensive examples (including SUT) to teach you how to

- Create a system behavioral model
- Set test coverage on the model
- Generate tests
- Execute tests

In practice, this means that the tool helps you to get an end-to-end model-based testing environment up and running.

The typical model-based testing workflow with TestCast MBT is as follows:

1. SUT behavior is modeled as UML state chart (tool has built-in model editor):
   a. External data definitions are in TTCN-3.
   b. EFSMs are used for modeling.
   c. Models must be deterministic models; if behavior is not deterministic there are techniques showing how to create a deterministic model.

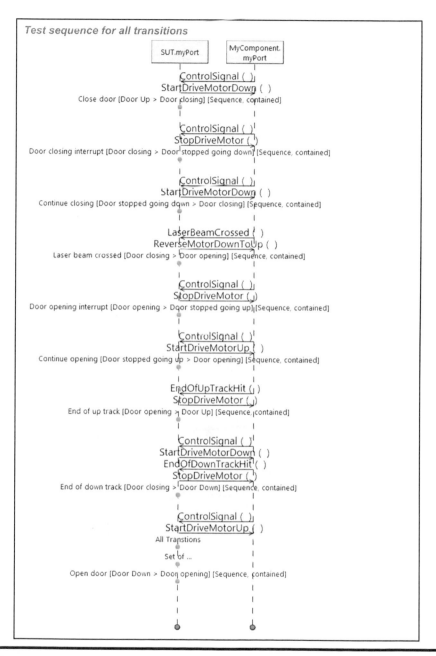

Figure 17.12   Derived (UML) sequence diagram.

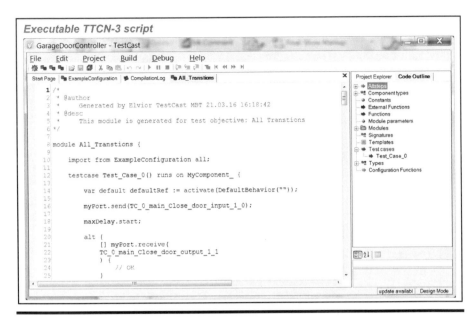

Figure 17.13   Partial TTCN-3 executable test script.

2. Structural test coverage criteria are set in terms of the model elements:
   a. List of transitions to be visited for example
   b. Predefined coverage such as "all transitions," "all transition pairs," and "all contained transitions"—in cases of hierarchical state machines
3. Abstract test sequences are automatically designed and generated:
   a. Test generation server is accessible via the Internet.
   b. Tests are designed automatically—shortest path to cover required criteria.
   c. Generated abstract test cases have required coverage.
4. Abstract tests are rendered into executable TTCN-3 scripts:
   a. After rendering, tests are ready for execution.
   b. TestCast MBT includes TTCN-3 tool (TestCast T3) for execution of TTCN-3 scripts.

Elvior provides TestCast evaluation licenses and licenses for academic use free of charge.

Chapter 18

sepp.med GmbH

18.1 Introduction

18.1.1 About sepp.med

Most of the material in this chapter was provided either by staff at sepp.med GmbH, or taken (with permission) from the tool website (www.mbtsuite.com). sepp.med is a medium-sized, owner-managed company, which specializes in IT solutions with integrated quality assurance in complex, safety-critical areas. sepp. med has its headquarters near Nuremberg (Germany), and other offices are located in Berlin, Wolfsburg, and Ingolstadt.

As a service provider, sepp.med focuses on the realization of complete services, which may be technical (requirements engineering, development, consulting, quality assurance, and test) as well as structural (project, process, and quality management). Beyond that, sepp.med is a well-reputed, certified training provider. For more information, please refer to the company's website www.seppmed.de (in German). During the past decades, sepp.med performed many quality assurance projects using model-based testing and jointly developed the automatic test case generator "MBTsuite" together with a partner, AFRA GmbH. Information on MBTsuite in English is available on www.mbtsuite.com, where there is also a contact form. sepp.med actively participated in the elaboration of the ISTQB® Certified Model-Based Tester training scheme. Senior consultant Dr. Anne Kramer is co-author of the related training book *Model-Based Testing Essentials– Guide to the ISTQB Certified Model-Based Tester: Foundation Level* [Kramer and Legeard 2016].

18.1.2 About MBTsuite

MBTsuite automatically generates executable test cases and test data from graphical models. The tool is designed for seamless integration into a variety of existing test processes. Instead of providing just another UML-like model editor, MBTsuite interfaces with various existing modeling tools. To ease the handling of the modeling tool and to support MBT-specific modeling, sepp.med developed a plug-in tool called "MBTassist," which provides a user interface to enter any parameter required by MBTsuite. It is available for Sparx Systems, Enterprise Architect, and IBM Rational Rhapsody.

MBTsuite takes a graphical model as input and generates executable test cases based on the chosen test selection criteria. The generated test cases and/or test scripts may in turn be transferred to an existing test management tool. Thus, MBTsuite bridges the gap between modeling, test management, and test automation tools, strictly focusing on the essential part of MBT, that is, test case generation from models (Figure 18.1).

To drive test case generation, MBTsuite implements a large variety of test selection criteria. By combining test generation parameters and subsequent filtering mechanisms, MBTsuite users are able to guide test case selection in a sophisticated way, thus obtaining a minimum set of test cases that fit the targeted test objectives best. It is even possible to define selection strategies individually for subdiagrams. Possible test case generation strategies are

- Full path coverage
- Random/stochastic test case selection
- Scenario-based selection called Named Path (one test case) and Guided Path (possibly several test cases)
- Shortest path, a project-driven test selection criterion based on a given weight type (cost, duration, or number of test steps)

MBTsuite implements two categories of filters, that is, coverage-based filters and range filters. Possible coverage-based filters are

- Requirements coverage
- Node coverage

Figure 18.1   MBTsuite integrates into existing tools.

- Edge coverage
- Test step/verification point coverage (based on a specific tagging of nodes and/or edges)

Range filters correspond to project-driven test selection criteria. Possible filter ranges are cost, duration, and weight. Obviously, those filters are only available, if the corresponding information has been included in the MBT model. Beyond the definition of test coverage statistics, MBTsuite offers various additional features to support test management, such as requirements traceability, prioritization, and visualization of specific test cases in the model.

18.1.3 Customer Support

MBTsuite is delivered together with comprehensive documentation, including a user guide, specific modeling guidelines, and dedicated profiles for some modeling tools. In addition, beginners may attend webinars at no charge. The MBTsuite service contract includes support in case of technical issues. For absolute MBT newcomers, sepp.med usually recommends the ISTQB® Certified Model-Based Tester training, which provides an overview on various MBT approaches. For MBTsuite users, standard and customized training courses, as well as coaching offers exist.

18.2 Insurance Premium Results

For the example in this chapter, we used Sparx Systems Enterprise Architect as the modeling tool, MBTsuite as the test case generator and MS Word as the output format. Other possible modeling tools are MID Innovator, IBM Rational Rhapsody, Artisan Studio, and Microsoft Visio. Regarding the output format, there is virtually no limitation including Application Lifecycle Management tools (HP ALM, MS Team Foundation Server, Polarion, …), test management tools, and test automation frameworks (NI TestStand, Ranorex, Selenium, Vector Cast, Parasoft, … and script languages (Java, .Net, Python, …). We decided to model the insurance premium example as UML activity diagram.

18.2.1 Problem Input

To write a test specification, we have to first analyze the requirements. Model-based testing is no exception from this rule. For traceability reasons, we formalized the requirements as shown in Table 18.1. The first column contains the requirement key, a unique identifier we will reference in our model later.

In this simple example, we created the requirement keys manually in the modeling tool. For larger requirements specifications, it is possible to import them automatically via a CSV file.

Table 18.1 **Insurance Premium Requirements (Identifier and Description)**

Requirement Key	Description
req_base_rate	The base rate is $600.
req_age_limits	People under age 16 or over 90 cannot be insured.
req_age_ranges	The following premium multiplication values for age ranges apply:

Age ranges	Age multiplier
16 <= age < 25	$x = 1.5$
25 <= age < 65	$x = 1.0$
65 <= age < 90	$x = 1.2$

req_fault_claims	The following premium penalty values at fault claims apply:

At Fault claims in past five years	Claims penalty
0	$0
1–3	$100
4–10	$300

req_fault_claim_limit	Drivers with more than 10 at fault claims in the past five years cannot be insured.
req_non_drinker_reduction	The reduction for being a nonDrinker is $75.
req_student_reduction	The reduction for being a goodStudent is $50.

The tests are based on the workflow of an insurance company employee working with the software under test. Obviously, in this example, he or she had to make some assumptions. In reality, this early modeling phase is characterized by intensive discussions with stakeholders, during which hidden or unclear requirements are clarified.

Figure 18.2 shows the resulting main diagram. It illustrates the most important modeling elements. In particular, we distinguish between test steps and verification points (abbreviated as: VP). Test steps are actions performed by the tester to stimulate the system under test. Verification points describe actions performed in order to observe and assess the reaction or output of the system under test. Figure 18.2 provides a high-level view of the workflow. Usually, it is possible to write this diagram at an early stage of the development process, because it is not yet necessary to

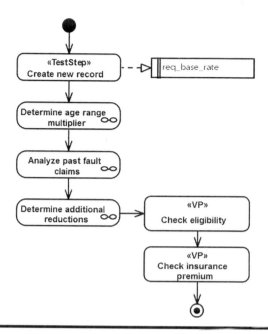

Figure 18.2 Insurance premium—main diagram.

know the implementation details. In a top–down modeling approach, we will then refine the activities in subdiagrams (indicated by the infinity sign in Figure 18.2).

Figure 18.3 shows the first subdiagram "determine age range multiplier," where we modeled the equivalence classes defined in the requirements. As you can see, we referenced the corresponding requirement keys in the diagram to facilitate model review and to enable automated requirement traceability.

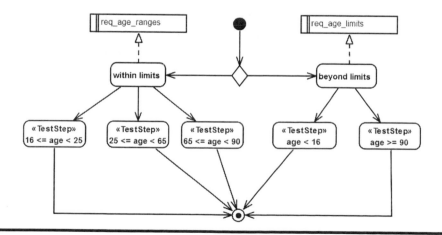

Figure 18.3 Insurance premium—"determine age range multiplier" (subdiagram).

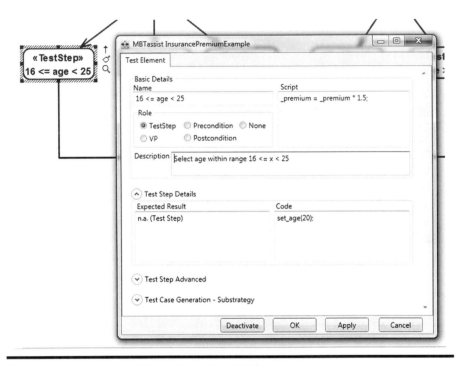

Figure 18.4   Entering information using MBTassist.

The diagram itself visualizes only part of the information contained in the model. Figure 18.4 shows the MBTassist input dialog. Please note that we compute the test oracle, in this case the insurance premium in dollars (see edit field "Script" to the top right), for later use. The "Description" field contains instructions for manual test execution; the "Code" field contains keywords or code snippets for automated test execution. Note the slight difference between the description and the code field: unlike in the code field, the description field does not contain a concrete value for the person's age. This is only for pedagogical reasons to illustrate the generation of both abstract and concrete tests.

Figures 18.5 and 18.6 show the remaining two diagrams "Analyze past fault claims" and "Determine additional reductions." To avoid creating test cases with past claims for persons that are not eligible, anyway, we use guard conditions. The value of the variable "_eligible" is set in the two nodes "within limits" and "beyond limits" in Figure 18.3, as well as in "below limit" and "above limit" in Figure 18.5.

The two variables "_eligible" and "_premium" are our test oracles. We use them again in the verification points in Figure 18.2. Figure 18.7 shows the MBTassist input dialog for the verification point "Check insurance premium." In the test case generation step, MBTsuite computes the value and finally replaces the variable ${_premium} by its value. In fact, MBTsuite integrates a Python script interpreter, which allows the user to perform complex operations and even to read data from external sources during test case generation.

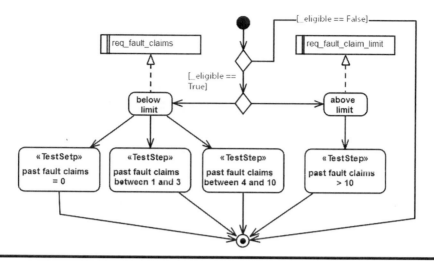

Figure 18.5   Insurance premium—"analyze past fault claims" (subdiagram).

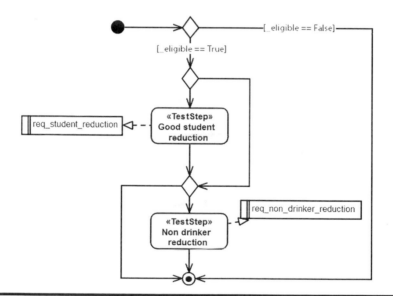

Figure 18.6   Insurance premium—"determine additional reductions" (subdiagram).

18.2.2 Generated Test Cases

To generate the test cases, we export the model from the modeling tool and import it into MBTsuite. The exact export/import mechanism depends on the modeling tool. For Sparx Systems Enterprise Architect, we use the built-in XMI export.

Once the model has been imported, we generate our test cases by applying the "Full path coverage" generation strategy. As a result, we obtain 41 test cases

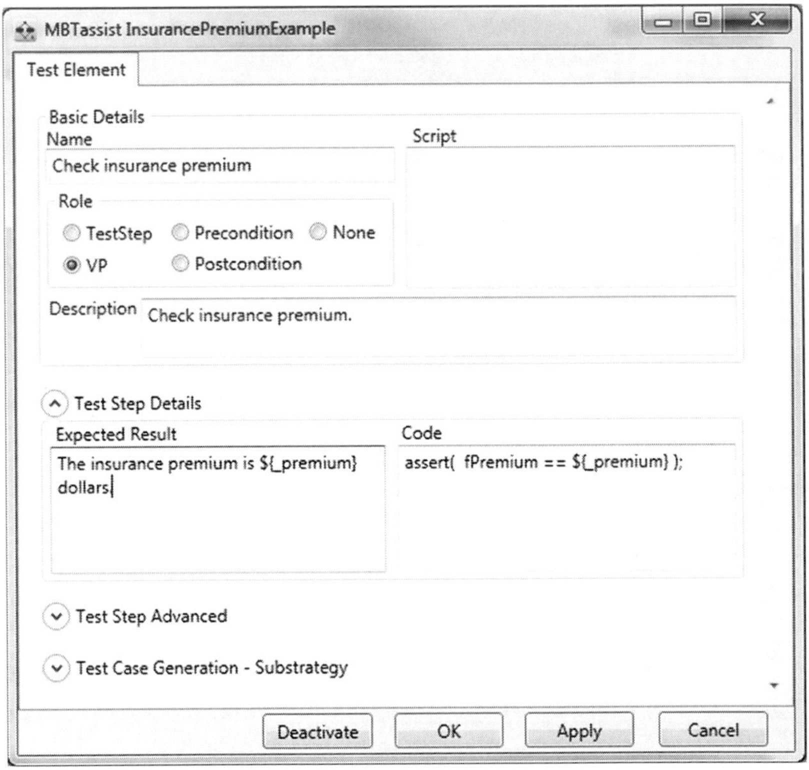

Figure 18.7 Using variables in MBTassist.

(indicated in the "Log" console as well as in the "Statistics" field in Figure 18.8). Those test cases correspond to the 39 rules in the decision tables in Section 13.2.2.2 but for one exception. We obtained three test cases for rule 39, corresponding to the three valid equivalence classes for the age range.

Usually, there is no time for exhaustive testing. Therefore, we have to reduce the number of test cases to a reasonable amount, but we still wish to ensure a certain degree of quality. Using MBTsuite, we apply additional filter criteria on those 41 test cases. In Section 18.2.2.1, we will present the six test cases obtained by applying the edge coverage filter.

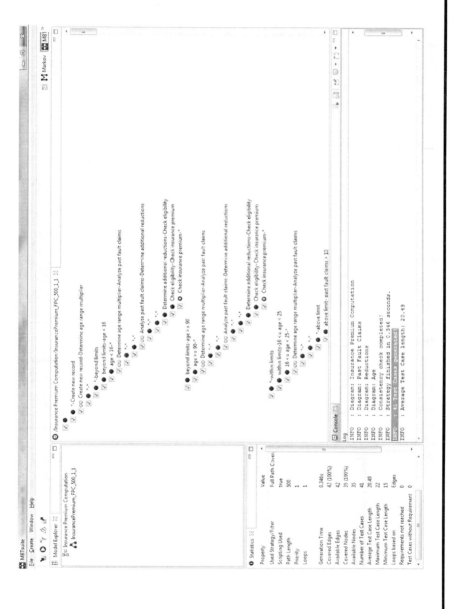

Figure 18.8 MBTsuite—generated test cases.

18.2.2.1 Abstract Test Cases Generated by MBTsuite, Activity Diagram "Insurance Premium"

Test Case 1: Premium_adult
Description
Age within limits past fault claims below limit
Requirements

- ◼ req_age_ranges
- ◼ req_base_rate
- ◼ req_fault_claims

This level of detail is provided only for Test Case 1.

Step	Type	Step Name	Step Description	Expected Result	Passed/ Failed
1	Test step	Create new record	Create new insurance record with a base rate of 600 dollars	n.a. (Test step)	☐/☐
2	Test step	25 <= age < 65	Select age within range 25 <= x < 65	n.a. (Test step)	☐/☐
3	VP	Check eligibility	Check whether person can be insured	True	☐/☐
4	VP	Check insurance premium	Check insurance premium	The insurance premium is 600.0 dollars	☐/☐

Test Case 2: Premium_child
Description
Age beyond limits
Requirements

- ◼ req_age_limits
- ◼ req_base_rate

Test Case 3: Premium_elder
Description
Age beyond limits
Requirements

- ◼ req_age_limits
- ◼ req_base_rate

Test Case 4: Premium_junior
Description
Age within limits, past fault claims below limit, nonDrinker
Requirements

- req_age_ranges
- req_base_rate
- req_fault_claims
- req_non_drinker_reduction

Test Case 5: Premium_junior0002
Description
Age within limits, past fault claims below limit, goodStudent nonDrinker
Requirements

- req_age_ranges
- req_base_rate
- req_fault_claims
- req_non_drinker_reduction
- req_student_reduction

Test Case 6: Premium_senior
Description
Age within limits, past fault claims above limit
Requirements

- req_age_ranges
- req_basc_rate
- req_fault_claim_limit

The six generated test cases are summarized in Table 18.2.

Table 18.2   Six Generated Test Cases

Test Case	Age	Claims	goodStudent	nonDrinker	Output
1	25 <= age < 65				Eligible
2	age < 16				Not eligible
3	age >= 90				Not eligible
4	25 <= age < 65	4 <= claims <= 10		T	$1125
5	25<= age < 65	1 <= claims <= 3	T	T	$875
6	65 <= age < 90	claims > 10			Not eligible

18.2.2.2 Concrete Test Cases Generated by MBTsuite, Activity Diagram "Insurance Premium"

The test cases presented in the previous section are abstract tests, because they did not contain any concrete value for the age and the number of claims. In addition, we illustrated the example of generating a test procedure specification for manual test execution. In this section, we address automated test execution of concrete test cases. Again, the distinction is for pedagogical reasons only. In reality, we would probably model concrete tests for manual execution too.

To obtain the test script in Figure 18.9, we used a different export format; in this case, the code exporter. The generated test case is a concrete test case, since it contains values both for the person's age and the number of claims.

In fact, we have several possibilities to turn abstract test cases into concrete test cases:

1. We use concrete values in the description, as we did in the "Code" field in Figure 18.4.
2. We model the boundary values, as we did for the equivalence classes in Figure 18.3.
3. We read the concrete values from external source using the Python interpreter (during test case generation).
4. We manage the concrete values in a test management tool.

Obviously, the fourth alternative is always possible and beyond scope of MBTsuite.

```
23    void Premium_adult(void)
24    {
25
26        /* Create new insurance record with a base rate of 600 dollars. */
27        create_record();
28        /* Expected Result: n.a. (Test Step) */
29
30        /* Select age within range 25 <= x < 65 */
31        set_age(45);
32        /* Expected Result: n.a. (Test Step) */
33
34        /* Set number of past fault claims to 0 */
35        set_claims(0);
36        /* Expected Result: n.a. (Test Step) */
37
38        /* Check, whether person can be insured. */
39        assert( bEligible == True );
40        /* Expected Result: True */
41
42        /* Check insurance premium. */
43        assert( fPremium == 600.0 );
44        /* Expected Result: The insurance premium is 600.0 dollars. */
45
46    }
```

Figure 18.9 MBTsuite—generated test script.

18.2.3 Other Vendor-Provided Analysis

We cannot present all features of MBTsuite in this chapter. The attentive reader might have recognized that both the test case name and the test case description were generated automatically based on advanced settings in the model. Moreover, there are many other possible filters and export formats available. For example, it is possible to generate a traceability matrix listing all requirements and the associated test cases. Last, but not least, it is possible to compare the generated test cases after having changed the model to the initially generated set of test cases prior to the model change. Using the so-called "Delta Tree Generation," you can determine whether the generated test cases are new, unchanged, changed, or obsolete.

18.3 Garage Door Controller Results

In the insurance premium example, we based the test case generation on activity diagrams. The Garage Door Controller example is different, leading to a different choice regarding the modeling notation, that is, a state diagram. Otherwise, the tool chain is identical to the previous example.

18.3.1 Problem Input

Once again, we start with a detailed requirements analysis, leading to the list of formalized requirements in Table 18.3.

We assume the existence of a testing framework, which allows us to send the four different control signals:

1. The controller device signal—control_signal
2. The end of up track signal—up_track_signal
3. The end of down track signal—down_track_signal
4. The safety device signal—safety_device_signal

Figure 18.10 shows the state diagram for the Garage Door Controller. You will notice two new elements, that are the pre- and postconditions, which we had not used in the previous example to keep it less confusing. MBTsuite collects all pre and postconditions defined in the model on the test cases path. Thus, we are able to define the preconditions in a subdiagram where they logically belong and, nevertheless, obtain a concatenated summary preceding the generated test case.

In Figure 18.10, we modeled our tests in a way that they stop with a closed door. This is an arbitrary choice. We could as well have decided to stop with an open door or to allow both choices as we did for the start.

From an MBTsuite point of view, the major difference between state diagrams and activity diagrams is the edges, that is, the arrows in the diagram. In activity

Table 18.3 Garage Door Controller Requirements (Identifier and Description)

Requirement Key	Description
req_close	When the door is fully open, and a signal from the control device occurs, the door starts closing.
req_open	When the door is fully closed, and a signal from the control device occurs, the door starts opening.
req_stop_manually	When the door is in motion, either closing or opening, and a signal from the control device occurs, the door stops.
req_resume	A subsequent control signal starts the door in the same direction as when it was stopped.
req_stop_end_of_up	When the door has moved to the extreme upper position (fully open), and a signal from the End of Up track occurs, the door stops.
req_stop_end_of_down	When the door has moved to the extreme lower position (fully closed), and a signal from the End of Down track occurs, the door stops.
hz_reverse	While the door is closing, if either the light beam is interrupted (possibly by a pet) or if the door encounters an obstacle, the door immediately stops and then reverses direction.

diagrams, they indicate a flow, which is possibly restricted by guard conditions. We may also name the edges to drive scenario-based test selection from activity diagrams. For all other information, MBTsuite analyzes actions and activities only.

In state diagrams, the transitions are much more important. Consequently, it is possible to model them as test steps or verification points. In Figure 18.10, nearly all transitions represent test steps and nearly all states are verification points. We just hid the corresponding tag to improve readability. There is only one limitation due to the modeling tool: it is not possible to link requirements to transitions.

Again, we used the Python scripting functionality of MBTsuite. Since finite state machines do not have a memory, the outcome of state "Stopped" is nondeterministic unless MBTsuite memorizes whether the door is opening or closing. Therefore, we introduce a variable called "_direction." Its value is set to UP in state "Opening" (as you can see in the field "Script" to the top right of Figure 18.11) and to DOWN in state "Closing." We use the variable in the guard conditions on the transitions between "Stopped" and "Opening" (_direction == "UP") and "Closing" (_direction == "DOWN").

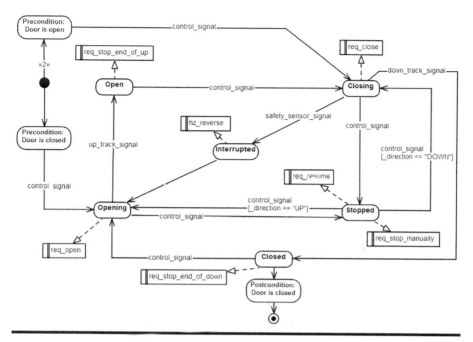

Figure 18.10 Garage Door Controller—state diagram.

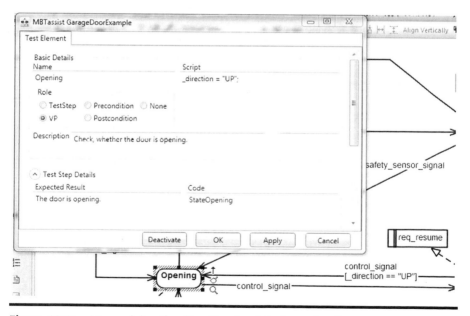

Figure 18.11 Memorizing the direction in state "opening."

18.3.2 Generated Test Cases

The basic procedure for test case generation is completely identical to the one described in the previous example. The test selection criteria and export formats presented in this section also apply to activity diagrams and vice versa. To generate our tests, we start again with full path coverage. However, we have to be very careful, because state diagrams are much more prone to test case explosion than are activity diagrams. In the insurance premium example, the diagram had no loop at all. In the Garage Door Controller example, we have to limit the number of loops taken into account at test case generation. By default, MBTsuite considers any path that traverses the same transition twice to be a loop, but the MBTsuite user may change this configuration.

With full path coverage and a loop depth of 1, we obtain 300 possible test cases. Now, we would like to reduce this number based on prioritization. First, we do not necessarily have to start both with a closed and with an open door. In the model in Figure 18.10, we may therefore assign a lower priority to the path between the start point and the precondition "Door is open." The priority is visualized in Figure 18.10 by two angle brackets "<<2>>," where 2 is the priority. At test case generation, we are able to limit the generated test cases to paths with priority equal or higher than the indicated value. (By default, all transitions have priority 0, which is the highest available value.)

With loop depth of 1 and priority 1 or higher, we still obtain 81 test cases. To limit them further, we may use the so-called "weight filter." The basic idea is to give a higher weight to some nodes or edges than to others. This enables MBTsuite to compute the total weight of each generated test case and to filter on this value.

Figure 18.12 shows the MBTassist user interface for adding the weight information to the model, whereas Figure 18.13 shows the MBTsuite user interface for

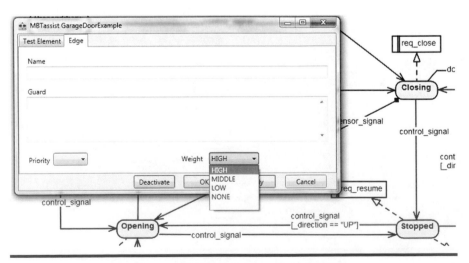

Figure 18.12   Adding data for the weight filter with MBTassist.

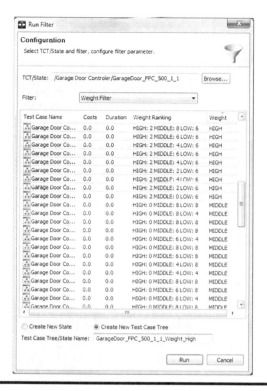

Figure 18.13   MBTsuite—weight filter.

test case selection based on the weight filter. It is possible to select tests based on their total weight (last column) or on the weight ranking. Other possible criteria are costs or duration, if the information has been added previously to the model (which is not the case in this example).

By limiting the priorities and using the weight filter, it is possible to steer test case selection in a very sophisticated way. Obviously, we could also use "full edge coverage," which brings us down to two generated test cases.

18.3.2.1 Abstract Test Cases Generated by MBTsuite, State Diagram "Garage Door Controller"

An abstract test case generated by MBTsuite from the model in Figure 18.10 could be a list of keywords for keyword-driven testing as illustrated in Figure 18.14. The underlying mechanism is the same as for activity diagrams. We have to include the keywords in the field "Code" and use an appropriate export format. Figure 18.14 is just an excerpt from a larger list of test cases. The exact export format is configurable.

```
SignalControlDevice
StateOpening
SignalControlDevice
StateStopped
SignalControlDevice
StateOpening
SignalControlDevice
StateOpen
SignalControlDevice
StateClosing
SignalControlDevice
StateStopped
SignalControlDevice
StateClosing
StateInterrupted
StateOpening
StateOpen
StateClosing
StateClosed
```

Figure 18.14 MBTsuite—generated keyword list (simplified example).

18.3.2.2 Concrete Test Cases Generated by MBTsuite, State Diagram "Garage Door Controller"

The concrete test case presented in this section is again for manual test execution. Unlike in the insurance premium example, where we presented the MS Word export format, we used the MS Excel export instead. Table 18.4 is a sample test case in spreadsheet format.

Table 18.4 Generated Test Case in Spreadsheet Form

Test Case	*Garage Door Controller*
Requirements	req_close req_stop_manually req_resume hz_reverse req_open req_stop_end_of_up req_stop_end_of_down
Precondition(s)	The door is open.
Postcondition(s)	The door is closed.

(Continued)

Table 18.4 (*Continued*) Generated Test Case in Spreadsheet Form

Step	Type	Step Name	Step Description	Expected Result	Requirements
1	Test step		Send control device signal.		
2	Verification point	Closing	Check whether the door is closing.	The door is closing.	req_close
3	Test step		Send control device signal.		
4	Verification point	Stopped	Check whether the door has stopped.	The door has stopped.	req_stop_ manually req_resume
5	Test step		Send control device signal.		
6	Verification point	Closing	Check whether the door is closing.	The door is closing.	req_close
7	Test step		Send control device signal.		
8	Verification point	Stopped	Check whether the door has stopped.	The door has stopped.	req_stop_ manually req_resume
9	Test step		Send control device signal.		
10	Verification point	Closing	Check whether the door is closing.	The door is closing.	req_close
11	Test step		Send safety sensor signal.		
12	Verification point	Interrupted	Check whether door is still moving.	The door has stopped.	hz_reverse
13	Verification point	Opening	Check whether the door is opening.	The door is opening.	req_open
14	Test step		Send signal "end_of_up".		

(*Continued*)

Table 18.4 (*Continued*)   Generated Test Case in Spreadsheet Form

Step	Type	Step Name	Step Description	Expected Result	Requirements
15	Verification point	Open	Check whether the door is open.	The door is open.	req_stop_ end_of_up
16	Test step		Send control device signal.		
17	Verification point	Closing	Check whether the door is closing.	The door is closing.	req_close
18	Test step		Send safety sensor signal.		
19	Verification point	Interrupted	Check whether door is still moving.	The door has stopped.	hz_reverse
20	Verification point	Opening	Check whether the door is opening.	The door is opening.	req_open
21	Test step		Send control device signal.		
22	Verification point	Stopped	Check whether the door has stopped.	The door has stopped.	req_stop_ manually req_resume
23	Test step		Send control device signal.		
24	Verification point	Opening	Check whether the door is opening.	The door is opening.	req_open
25	Test step		Send control device signal.		
26	Verification point	Stopped	Check whether the door has stopped.	The door has stopped.	req_stop_ manually req_resume
27	Test step		Send control device signal.		

(*Continued*)

Table 18.4 (*Continued*)   Generated Test Case in Spreadsheet Form

Step	Type	Step Name	Step Description	Expected Result	Requirements
28	Verification point	Opening	Check whether the door is opening.	The door is opening.	req_open
29	Test step		Send signal "end_of_up".		
30	Verification point	Open	Check whether the door is open.	The door is open.	req_stop_end_of_up
31	Test step		Send control device signal.		
32	Verification point	Closing	Check whether the door is closing.	The door is closing.	req_close
33	Test step		Send signal "end_of_down".		
34	Verification point	Closed	Check whether the door is closed.	The door is closed.	req_stop_end_of_down

18.3.3 Other Vendor-Provided Analysis

One of the major advantages of MBT is early requirement validation. In this example, we identified two missing requirements (see Table 18.5). One of them is quite straightforward, while the other still requires clarification.

If we consider the new requirements, the model in Figure 18.10 will become more complicated. Therefore, we recommend splitting the model into two parts:

Table 18.5   Garage Door Controller—Missing Requirements

hz_no_reverse	While the door is opening, if either the light beam is interrupted or if the door encounters an obstacle, the door continues to move in the same direction.
req_reverse_manually	Two subsequent control signals within (interval TBD) start the door in the opposite direction as when it was stopped.

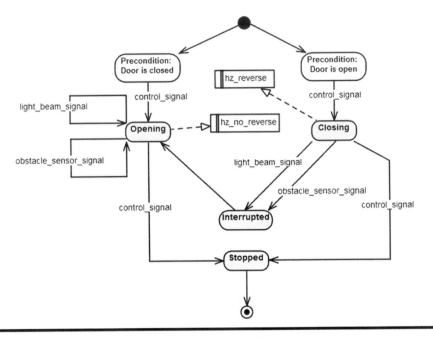

Figure 18.15   Garage Door Controller—testing the safety devices.

one for testing the normal use and one specifically for testing the safety devices. Splitting the model helps reducing complexity, because we may neglect all the aspects of normal use, just focusing on the safety devices. Remember: to keep it simple, we merged light beam sensor and obstacle sensor into one safety device. Figure 18.15 shows a possible model for testing both safety devices.

18.4 Vendor Advice

Like any other organizational change, introducing MBT should not be underestimated. It is much more than the tool—using MBTsuite is quite easy. It is the modeling activity which requires a new way of thinking, some training, and, last but not least, experience. Companies wishing to introduce MBT should consider attending the ISTQB® Certified Model-Based Tester training to gather a first overview on the different MBT approaches. Some of our customers test the approach with an "MBT prototype." They select a functionality of medium complexity and design test cases using both approaches in parallel: one team writes the tests "manually," following the traditional document-based approach. The other team (usually from sepp.med) writes the model and generates the test cases using MBTsuite. In the end, it is possible to compare the results regarding efficiency (number of hours needed), effectiveness (number of defects found), and quality (coverage reached). The results are usually in favor of MBT, especially if you take into account that the MBT team is usually not familiar with the system under test in the beginning.

Reference

[Kramer and Legeard, 2016]
Kramer, Anne and Bruno Legeard, *Model-Based Testing Essentials—Guide to the ISTQB Certified Model-Based Tester: Foundation Level*. John Wiley & Sons, Hoboken, NJ, 2016.

Chapter 19

Verified Systems International GmbH

19.1 Introduction

Verified Systems International GmbH (https://www.verified.de) was established in 1998 as a spin-off company of the University of Bremen, Germany. The company specializes in the verification and validation of safety-critical or business-critical embedded systems and cyber-physical systems. In this field, the company provides

- Services—from code inspections and reviews to testing campaigns and formal verification of models and code.
- Tools—for test automation, code analysis, and worst-case execution time analysis.
- Hardware-in-the-loop test equipment—high-performance test engines based on hard real-time technology, using clustered multicore systems.

Verified Systems main customers come from the avionics, railways, and automotive domains.

In this chapter, one of the Verified Systems main products—the RT-Tester test automation and analysis tool box—is applied to the Insurance Premium and Garage Door Controller case studies. In particular, the model-based testing component RTT-MBT is discussed, which is then utilized to automatically generate tests directly from the test model, requiring little manual effort to achieve high test strength.

19.1.1 The RT-Tester Tool Box

The RT-Tester test automation and analysis tool box comprises several tool components for testing and analyzing software and integrated hardware–software systems. Figure 19.1 gives an overview of the different components that constitute the RT-Tester tool box.

The RT-Tester core system RTT-CORE supports test procedure development using the RT-Tester Real-time Test Language (RTTL). The RTTL is embedded into C/C++ as a host language and provides dedicated commands for setting up multithreaded hard real-time testing environments and specifying tests. Importantly, RTTL provides communication mechanisms for exchanging data between simulation components, test oracles,* and the system under test (SUT) itself. Moreover, the RTTL contains commands for synchronization (e.g., suspending a thread until a logical data condition becomes true), evaluation of PASS/FAIL criteria, recording of test verdicts, and logging of the SUT reactions observed. The Real-time Test Language can be used on all test levels, from designing and performing unit tests on a host PC to creating and executing HW/SW integration tests or system

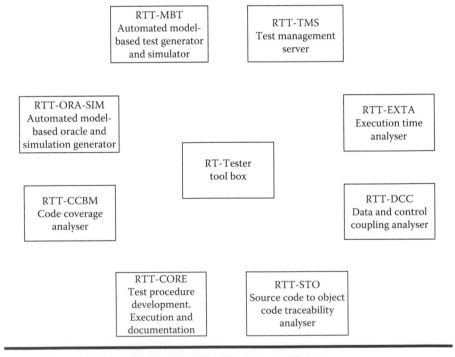

Figure 19.1 Main components of the RT-tester tool box.

---

* The checkers, which verify the behavior of the system under test with respect to the expected behavior, are frequently called test oracles.

integration tests of distributed embedded systems. The full power of the RTTL is typically exploited when designing HW/SW integration tests or system tests, where complex simulations of the actual operational environment are needed as a prerequisite for being able to stimulate the SUT reactions to be verified during a test.

The RT-Tester Test Management System RTT-TMS is used for managing multiuser testing processes, which can involve thousands of test cases that are implemented by large teams of test engineers. RTT-TMS enforces user-dependent roles and access rights, supports the delegation of testing tasks to test engineers, and provides extensive functions for evaluating product maturity and for reporting on the status of the test process. Finally, it produces traceability data that links requirements to test cases, showing where these test cases have been implemented in procedures, and associating the latter with the test results achieved. Requirements are usually imported from commercial requirements management systems such as IBM Rational DOORS.

The model-based testing component RTT-MBT is fully integrated with RTT-CORE and RTT-TMS, so that—instead of programming test procedures by hand—some or all test procedures to be created and executed in a testing campaign can be automatically generated from SysML models. This technology will be described in more detail in the next section.

RTT-ORA-SIM is a light-weight variant for model-based testing that provides a model-based code generator for creating environment simulations and test oracles from SysML models. Test inputs, however, are not directly derived from the test model. By way of comparison, RTT-MBT generates both the test inputs and the test oracles and can thus be seen as the more powerful model-based testing tool.

RTT-CCBM is a code-coverage analyzer (CCBM is an acronym for *code-coverage branch monitor*), which can be used to monitor the structural coverage achieved when executing tests on the target hardware. RTT-CCBM instruments the source code of the SUT to determine statements and branches covered during test executions on the target hardware. Upon test termination, the coverage data is transferred from the SUT to the test environment, and RTT-CCBM postprocesses the coverage data and generates a coverage report, which can be conveniently investigated in a web browser.

In addition to the aforementioned tool components, several special-purpose static analyzers are part of the RT-Tester universe.

The RT-Tester execution time analyzer RTT-EXTA estimates worst-case execution times for embedded HW/SW systems, as is required for safety-critical applications in the avionic and the automotive domains. RTT-EXTA combines static analysis and performance measurement results obtained during HW/SW integration testing to estimate the worst-case execution times of the analyzed software. The tool specifically focuses on end-to-end execution times, that is, the time which is required until a system reacts to a certain stimulation.

RTT-DCC automates the so-called *data and control coupling* (*DCC*) *coverage analysis*, which is required for the certification of avionic systems according to the RTCA DO-178B/C. The objective of DCC analysis is to check whether the integration tests sufficiently examine all data-based dependencies in the system as specified in the software design. DCC analysis is thus performed to verify completeness of testing activities. In particular, RTT-DCC correctly handles pointers and aliasing, which usually make manual analysis infeasible due to the sheer complexity of contemporary software. Moreover, RTT-DCC checks whether all interleaved access situations from concurrent processes to shared data resources have been exercised. Although the manual analysis is very time consuming, RTT-DCC automatically determines all necessary test sequences to be executed for obtaining full DCC coverage.

For avionics systems of the highest criticality, that is, *Design Assurance Level A* according to the RTCA DO-178B/C standards, the object code generated by the compiler has to be verified with respect to traceability to the C-code from which it had been generated. RTT-STO automates this so-called *source-code-to-object-code traceability analysis (STO-analysis)*. The tool detects all control flow in the object code that is not traceable to the original C-code. It uncovers all compiler-inserted calls to built-in library functions that are used, for example, by the compiler to perform certain arithmetic operations. RTT-STO checks whether variables are allocated by the compiler with the size that is required, and it checks by means of abstract interpretation techniques, whether all writes to memory addresses contained in registers use well-defined address values.

In the following sections, we will focus on RT-Tester's model-based testing component RTT-MBT and discuss how it can be applied to produce test suites for the case studies—Insurance Premium Problem and the Garage Door Controller.

19.1.2 The Model-Based Testing Component RTT-MBT

RTT-MBT supports the automated generation of test cases, concrete test data, and test procedures from UML/SysML models. The models describe the expected behavior of the SUT, its interfaces to the testing environment, and, optionally, simulations describing admissible behaviors of the operational environment. The latter restrict the concrete sequences of input data to the SUT to those that can actually happen during SUT operation.* For creating test models, a set of different tools can be used. Currently, RTT-MBT supports the following UML/SysML modeling tools as front-ends:

■ EnterpriseArchitect by Sparx Systems Ltd (http://www.sparxsystems.eu/start/home/)
■ IBM Rational Rhapsody (http://www-03.ibm.com/software/products/en/ratirhapfami)

---

* In the case studies treated here, no restrictions to the operational environment apply, so we will not describe environment simulations in more detail.

- Astah SysML by (http://astah.net/editions/sysml)
- PTC Integrity Modeler (http://www.ptc.com/model-based-systems-engineering/integrity-modeler)
- Papyrus (https://eclipse.org/papyrus/—this is a freely available open source Eclipse plugin)

RTT-MBT is connected to these tools based on XMI exports. Other UML/SysML modeling front-ends can also be supported. RTT-MBT can also be extended to generate test procedures for other test automation platforms than RT-Tester, so that RTT-MBT can be used in combination with other testing tools (see Figure 19.2).

Creating test models. To facilitate test model development, templates are provided for each of the modeling tools listed above. These templates implement the basic structure of a typical test model and can thus be extended for concrete instances. UML composite structure diagrams or SysML internal block diagrams can be used to perform a functional decomposition of complex SUTs. In the case studies treated here, each test model consists of a single component. The required behavior of the SUT is modeled by means of

- UML/SysML state machines.
- UML/SysML activity diagrams.
- Operations using simple C-style commands in their bodies.

The state machines may contain logical conditions of model variables (SysML value parameters) and timing conditions (e.g., "after 150 ms") as transition guards. Alternatively, state machine transitions may be triggered by events and generate

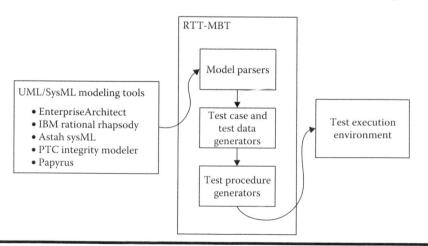

Figure 19.2 RTT-MBT architecture supporting different modeling front-ends and test execution environments.

events, as in Mealy automata. A test model consisting of several subcomponents can have several state machines (as in the Statechart notation), which are interpreted as concurrent components. Whenever transitions are enabled, each state machine then fires its transitions concurrently with the other state machines.[*]

Activity diagrams are used to model the expected behavior of SUTs consisting of small single-threaded software units, such as the one to be tested in the Insurance Premium calculation case study.

19.1.2.1 Test Case Generation

Test cases are automatically generated from the model, and a combination of different strategies may be used to steer the test case generation process. Each test case is internally represented as a logical formula, which characterizes the elements of the test model that shall be covered. For example, such a formula may characterize a specific MC/condition that needs to be examined by a test. MC/DC (multiple condition/decision coverage) is mandated for avionics systems. RTT-MBT then generates a sequence of test inputs such that the desired MC/DC condition is examined. For creating the concrete test data, a mathematical constraint solver in combination with heuristic search algorithms identifies a solution to each test case formula, from which the test case is derived. The solution is consistent with the model in the following sense: the solution must represent a valid path through the model, and at the same time, it has to fulfill the test case formula. We will illustrate this concept below when describing the Insurance Premium calculation case study.

RTT-MBT has built-in support for several test case generation strategies. The first class of these strategies extracts the test goals from the model by means of a syntactic model analysis. Such syntactic goals include state coverage, transition coverage, condition coverage, and combinations thereof. Further, RTT-MBT can automatically determine the equivalence classes implemented by a model and can thus be used to automatically achieve a high degree of test strength. This is based on a semantic analysis: the model behavior is encoded by means of a logical formula called *transition relation*. This formula relates model prestates to their possible poststates. With the transition relation at hand, equivalence classes of model states and model inputs can be identified. Finally, RTT-MBT can also be used to automate *conformance tests* establishing some correctness relation between the SUT and the reference model. It is important to emphasize that typically, test cases are derived using combinations of these strategies, which can be configured via a graphical user interface. For example, it is possible to verify some SUT properties against a functional subcomponent of the model by means of conformance testing, while

---

[*] In http://www.mbt-benchmarks.org more complex models have been published showing different variants how concurrency and timing conditions can be specified for automated model-based testing with RTT-MBT. For the case studies described here, neither concurrency nor timing conditions apply.

the overall functionality of the system is tested by means of equivalence class and model coverage test cases.

Overall, RTT-MBT provides the following test case generation strategies:

1. *Basic control state test cases* (*BCS*): Every simple state in a UML/SysML state machine gives rise to a test case specifying the goal that inputs to the SUT should be created such that this state is reached ("covered").

2. *Transition test cases* (*TR*): Every state machine transition gives rise to a test case whose goal is to execute this transition. This requires finding input sequences to the SUT such that first the transition's source state is reached, and then its guard condition becomes true, or its trigger event is provided by the testing environment, respectively.

3. *MC/DC test cases* (*MCDC*): Transitions involving composite guard conditions can be tested with different valuations of the guards atoms. The MC/DC coverage criterion for guard condition (a or b), for example, requires to test this transition with (a = true and b = false) and once with (a = false and b = true), so that first a causes the transition to fire (because b is false), and then b causes it to fire. RTT-MBT also creates a *stability test case* for this situation, where both a and b evaluate to false. It is then checked that the transition does *not* fire.

4. *Hierarchic control state test cases* (*HITR*): UML/SysML state machines can have hierarchic composite states (so-called OR-states), where a super state is decomposed into a lower level state machine (so-called submachine). The lower level state machine is executed as long as the whole state machine resides in the super state. When testing a high-level transition emanating from the super state, the lower level state machine may reside in differing lower level states. Therefore, different test cases are generated for this higher level transition, one for each lower level state the submachine may reside in.

5. *Basic control state pairs test cases* (*BCSPAIRS*): For test models involving several concurrent state machines, RTT-MBT identifies test cases for each pair of states in interacting state machines. If, for example, state machines SM1 and SM2 interact, and SM1 has simple states s11, s12, s13, while SM2 has simple states s21, s22, then RTT-MBT specifies test case conditions such that model state combinations (s11,s21), (s11,s22), (s12,s21), …, (s13,s22) are targeted. RTT-MBT then generates test cases to cover these combinations.

6. *Equivalence class and boundary value test cases* (*ECBV*): RTT-MBT analyzes the input data types of a test model, as well as the guard conditions occurring in state machines or in activity charts. As a result of this analysis, it identifies input equivalence classes, such that different inputs from the same class will lead to equivalent model reactions (i.e., equivalent expected reactions of the SUT), whenever these inputs are applied to equivalent model states. The identification of these classes uses sophisticated analysis techniques that have been published [Huang and Peleska 2016]. The resulting test cases have significantly

higher test strength compared to conventional techniques [Hübner et al. 2015]. For the equivalence classes, the constraint solver integrated in RTT-MBT also calculates boundary value tests.

7. *Conformance test cases* (*CONF*): For models of moderate complexity or for subcomponents of large models, test cases proving conformance* can be generated. This means that a finite test suite is created that guarantees under certain hypotheses to uncover *every* deviation of the SUT from the UML/SysML reference model. These hypotheses refer to the maximal number of internal state equivalence classes that may occur in a possibly faulty implementation and to the granularity of the input equivalence classes chosen: a faulty SUT may behave correctly for a subpartition of an input equivalence class derived from the model, but act incorrectly in another subpartition. This problem is mitigated by refining the original input equivalence class partition and by selecting random values (including boundary values) from each input class, whenever a representative of that class is needed. Experiments have shown that even if a faulty SUT does not satisfy the fault hypotheses, the test strength of the resulting test suite is superior to heuristic or random test generation techniques [Hübner et al. 2015].

For the first case study, the equivalence class and boundary value test cases according to Item 6 will be applied, and for the second case study, a conformance test suite (Item 7) will be created.

19.1.2.2 Test Procedure Generation

RTT-MBT generates test scripts, which are then compiled into executable test procedures by the RTT-CORE system component of RT-Tester. Each procedure may execute one or more test cases of the different classes listed above against the SUT. During the test procedure generation process, the logical conditions defining each test case are internally solved by the constraint solver and by application of search heuristics. From the solution, a sequence of input vectors with associated timing conditions is extracted. This sequence then serves as input data to the SUT. The test oracles are generated by automatically transforming the model into executable checkers (test oracles) monitoring each output interface of the SUT. Each checker observes which input vectors are sent to the SUT, calculates the expected SUT reactions as specified by the model, and compares the expected reactions to the actual values observed on the output interfaces of the SUT during test execution.

---

* The current version of RTT-MBT supports the following notion of conformance: every input/output sequence that can be performed by the reference model can also be performed by the SUT, and every I/O sequence performed by the SUT is admissible according to the model. This relation is also called *I/O equivalence* or *language equivalence*.

The test engineers expertise is required for configuring the test procedure generation process by deciding which test cases should be exercised within the same procedure. For this configuration process, three approaches are supported.

1. *Requirements driven approach*: SysML models can link model elements— for example, the states and transitions of state machines—to requirements, using the so-called *satisfy-relationship* between model elements and requirements. If these relationships have been specified in the model, RTT-MBT will automatically relate requirements to test cases: a test case contributes to the verification of a certain requirement, if the test case covers (some or all of) the model elements linked to the requirement. Using the configuration panel shown in Figure 19.3, test engineers can simply select the requirement(s) to be tested by a test procedure by dragging the requirement tag(s) to the right-hand side of the panel. RTT-MBT automatically identifies all test cases linked to the selected requirements, generates the associated input test data to the SUT, and transforms input data and test oracles into a new test procedure which can be compiled and executed in the RTT-CORE system.

2. *Model coverage driven approach*: In the model-driven test procedure generation approach, the test engineers just select a portion of the model, such as

Figure 19.3 Configuration panel for requirements driven test procedure generation. A requirement is selected from the list of all requirements extracted from the model and dragged/dropped onto the right-hand side configuration panel for test procedure generation.

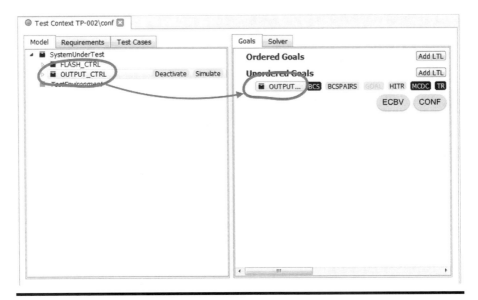

Figure 19.4 Test procedure generation is configured by selecting the model portion and the associated test case types to be covered. Here, block OUTPUT_CTRL is selected, to be tested by all applicable basic control state, transition, and MC/DC test cases.

a SysML block or a number of state machine transitions and drags/drops them to the configuration panel, as shown in Figure 19.4. Moreover, the engineers select the test strategies (BCS, TR, ..., ECBV, CONF) to be applied. RTT-MBT then generates a test procedure that realizes all test cases according to the selected configuration. This is shown in Figure 19.5. According to the selection shown there, all basic control state test cases, transition coverage, and MC/DC coverage test cases referring to states and transitions in block OUTPUT_CTRL will be realized in the test procedure to be created.

3. *Test case driven approach*: If no requirements have been identified, or if test cases linked to different requirements should be executed by the same procedure, the test case-driven approach to test procedure generation may be used. Here the test engineer selects the test cases in the test case selection panel and drags/drops them onto the configuration pane, as shown in Figure 19.5. The test case identifiers indicate which strategy has been applied for their generation: a "BCS" in the tag specifies that it has been generated to ensure the coverage of a certain state machine state, a "TR" indicates that it has been created to ensure coverage of a given state machine transition, and so on.

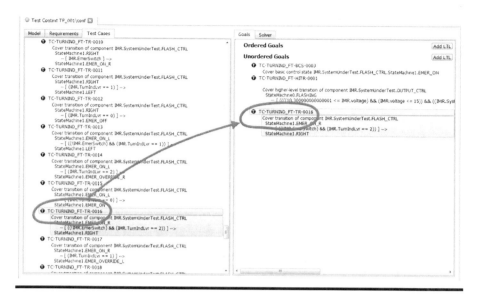

Figure 19.5 Test procedure creation is configured by selecting the test cases to be executed.

19.2 Case Study: Insurance Premium Calculation

For model-based unit testing, RTT-MBT uses activity diagrams as test models. Such a diagram is shown in Figure 19.6 for the insurance premium calculation case study. Since the input parameters Age and BaseRate have a wide range of admissible values, an exhaustive test using all possible parameter combinations is not advisable. Therefore, we select the equivalence class and boundary value test strategy for the block containing the activity diagram. As an additional parameter, we specify that 10 concrete test data sets should be derived from each input equivalence class (including boundary values), so that the test procedure generated by RTT-MBT performs 10*(number of classes) invocations of the unit under test (UUT). Each invocation applies a different input vector (BaseRate, Age, Claims, goodStudent, nonDrinker) selected from the input equivalence classes. When calculating the input equivalence classes, RTT-MBT detects that the activity diagram is free of unbounded loops; therefore, it is feasible to generate fine-grained equivalence classes such that all input vectors from a given class lead to the same output "modulo BaseRate": this means that when fixing a BaseRate value, all input vectors of the same class result in the same output value. Since input equivalence classes are calculated by RTT-MBT such that the SUT or UUT shows equivalent behavior for all members of a class, it is not surprising that each class is specified by a logical condition over *several* input parameters. Input BaseRate, however, never occurs in these conditions, since none of the guard conditions in the activity chart depend on parameter.

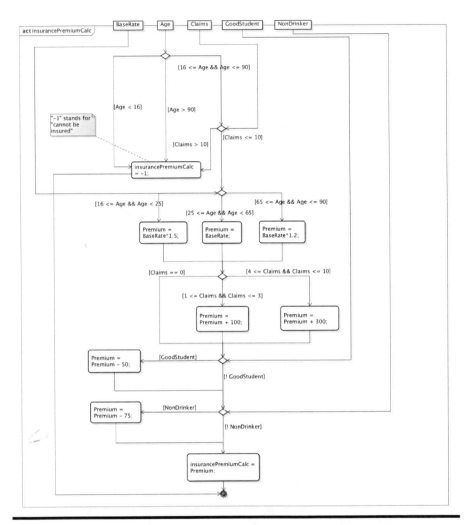

Figure 19.6 **Activity diagram specifying the insurance premium calculation.**

The input class specified by logical condition

(Age < 16) || (Age > 90) || (Claims > 10),

for example, comprises all input vectors (BaseRate, Age, Claims, goodStudent, non-Drinker), where Age is less than 16 or greater than 90, and Claims are greater than 10. Since the condition does not refer to BaseRate, goodStudent, or nonDrinker, this means that all possible value combinations of these input variables are combined with all Age and Claims values satisfying the above condition. This input equivalence class specifies the input conditions for the "cannot be insured" case, as can be easily seen in the activity chart by checking the entry conditions for action insurancePremiumCalc = -1;

which realizes the "cannot be insured" case.

For classes involving disjunctions in their logical conditions, the MC/DC rules are applied when selecting representatives. For the example above, at least one solution of each of the following subclasses is selected:

Age < 16 && Claims <= 10
Age > 90 && Claims <= 10
16 <= Age && Age <= 90 && Claims > 10

As another example, the equivalence class

(16 <= Age) && (Age < 25) && (1 <= Claims) && (Claims <= 3)
&&!goodStudent && nonDrinker

specifies the input partition where, for all inputs, the base rate is multiplied by 1.5, the at-fault claims increase the premium by an additional $100, and the non-Drinker reduction leads to a final decrease by $75. Summarizing, for all members of this class, the resulting insurance rate is *BaseRate**1.5 + 100 − 75. RTT-MBT calculates representatives of each class using its mathematical constraint solver. The solver could also handle arithmetic expressions in the equivalence class conditions, including floating point calculations and bit vector arithmetic for integers.

By transforming the logical conditions for each class, several boundary value conditions are created. For the above equivalence class, these are the conditions:

(Age == 24) && (1 == Claims) &&!goodStudent && nonDrinker
(Age == 16) && (1 == Claims) &&!goodStudent && nonDrinker
(Age == 24) && (3 == Claims) &&!goodStudent && nonDrinker
(Age == 16) && (1 == Claims) &&!goodStudent && nonDrinker

For expressions "$(x < y)$" with floating point numbers x, y, the boundary condition is specified by setting x to the smallest representable floating point value which is smaller than y.

When calculating the input equivalence classes, RTT-MBT enumerates all possible true/false values of Boolean parameters, as well as integers with a range up to 5. For variables with a wider range, equivalence class conditions such as the ones for Age and Claims shown in the example above are determined. In this case study, this leads to 37 input equivalence classes. As a consequence, the requirement "10 input vectors per class" results in 370 test cases, which also cover the boundary conditions described above. Selecting 10 representatives from each class helps to uncover errors in guard conditions implemented by the UUT, as well as arithmetic errors in the premium calculation which cannot be detected by a single input. Suppose, for example, that the UUT behavior is erroneous for the input class

(16 <= Age < 25) && (1 <= Claims <= 3) &&!goodStudent && nonDrinker,

Using the faulty formula *Premium* = *BaseRate* + 100 instead of *Premium* = *BaseRate**1.5 + 100 − 75. If we select just one input from this class, say,

(BaseRate, Age, Claims, goodStudent, nonDrinker) = (150, 24, 1, false, true),

the error would remain uncovered, since both correct and faulty formula yield the same value for this input vector. Using just one different value for BaseRate, however, the error would be detected.

It should be noted that the input equivalence class calculation method can also be applied to nonterminating state machines, and—under certain hypotheses—the resulting test suite can uncover every conformance violation of the SUT. The underlying theory has been described in [Huang and Peleska 2016].*

19.3 Case Study: Garage Door Controller

For the Garage Door Controller example, RTT-MBT supports finite state machines modeled in SysML style, as shown in Figure 19.7. The input signals e1 , ..., e4 are used as specified in the Garage Door Controller problem description, likewise the actions a1 , ..., a4. Additionally, the test cases generated by RTT-MBT will reference the *empty action* "—" which denotes that the system simply ignores a certain event in a certain state: recall that—at least for safety-critical or business-critical applications—it is also necessary to check the robustness of the SUT with respect to "unexpected" or "unwanted" events in a certain state. To this end, the standard SysML interpretation for state machines is to *ignore* trigger events not occurring on any transition emanating from a given state. In state Door_Up, for example, only one transition triggered by signal e1 is specified. The occurrence of events e2, e3, and e4 in this state do not have any effect. This behavior has to be tested, because it might be the case that the SUT is not sufficiently robust. The Garage Door Controller could potentially crash or perform some other unwanted action when receiving an unexpected signal—say, e2 (end of down track hit)—in state Door_Up, or analogously, in any other state.

RTT-MBT automatically analyzes the reference model for the expected SUT behavior. It detects that the model implements a deterministic finite state machine of manageable size and suggests applicable testing strategies accordingly, to be selected by the users.

---

* In 2015, Verified Systems International was awarded the runner-up trophy of the Innovation Radar Innovation Prize of the European Union for making this theory practically available the RTT-MBT product, see https://www.verified.de/publications/papers-2015/eu-innovation-radar-price-runner-up-trophy-for-verified-systems-international/

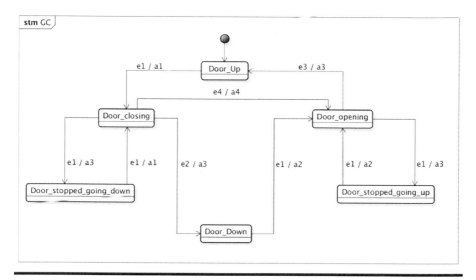

Figure 19.7   Garage Door Controller FSM.

- *Basic control state test cases*: This testing strategy results in a light-weight test suite, just enough test cases for visiting every state in the model.
- *Transition test cases*: The resulting test suite exercises each transition in the state machine model. However, it does not exercise the robustness situations in each state, which are concerned with the signals to be ignored.
- *Requirements coverage*: We could augment the model with a list of require-ments linked to elements of the reference model (i.e., the SysML state machine). Then, selective test suites could be generated for just testing a spe-cific requirement. This possibility is most suitable for regression testing, when the SUT has been modified to implement a changed requirement or an addi-tional one that has not been specified before.
- *Conformance testing*: This is the most thorough strategy. As described above, conformance testing requires tool users to indicate a *fault hypothesis*, which states how many internal states m the minimized state machine describing the true behavior of the SUT might have. The resulting test cases are sufficient to guarantee that *every* deviation of the SUT from the expected behavior will be detected, provided that the SUT behavior can be modeled by means of a deterministic state machine of at most m states. This strategy also exercises all robustness tests about signals to be ignored in any state.

Since the Garage Door Controller is mildly safety-critical (one certainly would not want a child or an animal to get stuck under the closing door), the conformance testing strategy is the one selected here.

First, the RTT-MBT tool minimizes the reference model. It turns out that the minimal deterministic finite state machine (DFSM), which is equivalent to the

SysML reference model in Figure 19.7, only has four states and can be represented as shown in Figure 19.8.[*] The input signals and output events are of course the same as the ones used in the SysML state machine shown in Figure 19.7. The states and transitions, however, differ as a consequence of the minimization process. The state labels, for example, "GC_MIN {0,2}(0)," signifies that the states with internal numbers 0,2 in the original SysML model have been collapsed into a single state with internal number 0 in the minimized state machine. The internal state numbers in the SysML reference model are 0 for Door_Up, 1 for Door_Down, 2 for Door_stopped_going_down, 3 for Door_stopped_going_up, 4 for Door_closing, and 5 for Door_opening. Therefore, the new finite state machine indicates that the original states Door_Up and Door_stopped_going_down, as well as Door_Down and Door_stopped_going_up are equivalent. This observation can straightforwardly be verified by analyzing the outgoing transitions and their target states for each of these states.

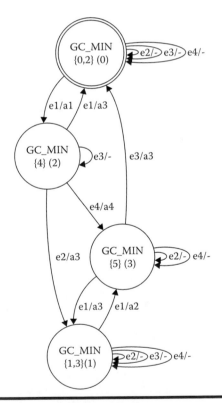

Figure 19.8 Minimized deterministic finite state machine equivalent to the reference model from Figure 19.7.

---

[*] RTT-MBT uses the Graphviz "*.dot"-format (http://www.graphviz.org) to output Graphs in various situations, such as for the finite state machine shown in Figure 19.8. Using the Graphviz tool, these graphs can be visualized and stored in various formats.

For example, we choose $m = 6$ for the maximal number of states occurring in the minimized finite state machine. RTT-MBT then generates 252 test cases, each case is described by a sequence of inputs and associated expected outputs derived from the reference model. For this calculation, the so-called Wp-Method is used [Luo et al. 1994]. Some examples of test cases are shown here:

1. (e1/a1).(e2/a3).(e1/a2).(e1/a3).(e1/a2).(e1/a3)
2. (e1/a1).(e2/a3).(e1/a2).(e1/a3).(e1/a2).(e2/-)
3. (e1/a1).(e2/a3).(e1/a2).(e1/a3).(e2/-).(e1/a2)
4. (e1/a1).(e2/a3).(e1/a2).(e1/a3).(e3/-).(e1/a2)
...
155. (e1/a1).(e4/a4).(e4/-).(e3/a3).(e1/a1).(e2/a3)
156. (e1/a1).(e4/a4).(e4/-).(e3/a3).(e2/-).(e1/a1)
157. (e1/a1).(e4/a4).(e4/-).(e3/a3).(e3/-).(e1/a1)
158. (e1/a1).(e4/a4).(e4/-).(e3/a3).(e4/-).(e1/a1)
159. (e1/a1).(e4/a4).(e4/-).(e4/-).(e1/a3).(e1/a2)
160. (e1/a1).(e4/a4).(e4/-).(e4/-).(e2/-).(e1/a3)
...
251. (e4/-).(e4/-).(e3/-).(e1/a1)
252. (e4/-).(e4/-).(e4/-).(e1/a1)

These test cases guarantee to find every fault in any SUT whose behavior is equivalent to that of a minimized DFSM with at most six states. Assume, for example, that the actual behavior of the SUT is reflected by the minimized DFSM shown in Figure 19.9. Its behavior conforms to the reference model in Figure 19.7, as long as it resides in states 0, 1, 2, 3. In state 3, however, the implementation contains a *transition fault* (this is also called a *trap door* [Binder 2000]): instead of ignoring input e4 (laser beam crossed), it transits to state 4. Since no visible output is produced in this step (the empty action "—" is invisible during test execution), this fault cannot immediately be detected. Further, when receiving the expected event e3 (end of up track hit), the SUT still responds correctly with action a3 (stop drive motor). A correct implementation would now show a behavior conforming to the Door_Up state. This SUT, however, has reached faulty state 5, where it produces a robustness failure on receiving event e4 (laser beam crossed). Instead of ignoring this event as is expected in the up-position, it produces action a1 (start drive motor down). This failure is quite subtle, because it can only be detected if

■ The test case is sufficiently long to reach the hidden faulty state 5.
■ After having reached state 5, a robustness test is performed with the event e4.

The equivalence test suite listed above produces test cases of sufficient length to reach every faulty hidden state, as long as the minimized DFSM representing the SUT has at most six states. Moreover, all input signals are exercised in every state; it is then checked whether the target state reached is equivalent to the one expected

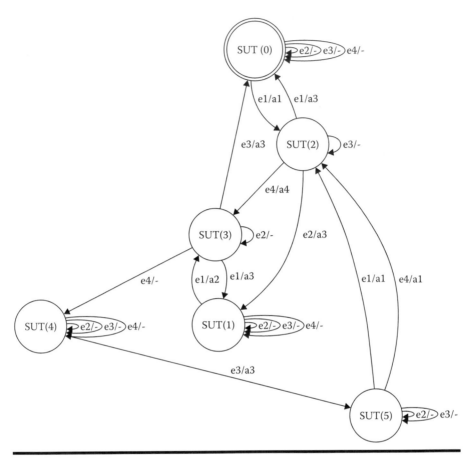

Figure 19.9   DFSM showing the true behavior of a faulty SUT.

according to the reference model. The faulty behavior described above is uncovered by test case

158. (e1/a1).(e4/a4).(e4/-).(e3/a3).(e4/-).(e1/a1),

because after the past e4-input, the empty action is expected, whereas the SUT produces the action a1. This example illustrates that the conformance test suite uncovers rather subtle implementation errors that probably would not be detected when test cases are designed based on intuition.

The choice of $m = 6$ is a good choice in this example, because the reference model already has six states in its original form. Hence, it appears very likely that an implementation has a similar number of states. The implementation, however, might be faulty, so that this number of states is already required for its minimized DFSM representation. The number of 252 test cases is quite acceptable if the test equipment allows for fully automated execution of these tests. Typical hardware-in-the-loop test configurations would simulate inputs e1, ..., e4 on the input interfaces of the Garage Door Controller and would check on-the-fly whether the responses of the controller conform to the ones expected according to the model. Even when executing these tests in realistic physical time, at most two minutes would be needed per test case, so the whole suite could be executed within 8.5 hours. With RT-Tester and the hardware-in-the-loop test equipment of Verified Systems International, this execution would be fully automated.

If, however, there is a high probability that the minimized DFSM reflecting the implementation behavior has four states only,* then a significantly shorter test suite could be derived: with $m = 4$, only 16 instead of 252 test cases would be needed. This test suite is still capable of detecting every output fault and every transition fault leading to one of the four "legal" states of the minimized reference model. It would, however, not be capable of detecting trap doors leading to *additional* faulty states.

When performing black-box tests, the proper value of m is not known. Simply increasing the size of m so as to include as many faulty behaviors as possible is not advisable because the number of test cases grows exponentially with m. For $m = 7$, for example, we would need 1010 test cases for the Garage Door Controller example. A reasonable approach to choosing m is therefore to select a number which is greater, but still rather close to the number of states in the minimized DFSM associated with the reference model. If this does not uncover any errors, additional random *long duration tests* should be executed to ensure that the SUT behavior conforms to that of the reference model over a long period of time.

19.4 Vendor Advice

At Verified Systems International GmbH, model-based testing projects performed with RTT-MBT for customers as a service usually yield an efficiency increase of 30% in comparison with conventional testing campaigns where test case identification, test data creation, and test procedure programming are performed in a manual way. This efficiency increase is achieved in the first campaign, where the test models still have to be created from scratch as part of the MBT activities. For regression-testing campaigns, the efficiency increase is significantly higher, since then only small updates need to be performed on the test models. At the same time, the quality of the tests is always better than that of manual test campaigns, since it is far easier to achieve a comprehensive coverage when this is created by a tool in an automated way than in the case where test cases, data, and procedures have to be elaborated manually.

---

* This could be assumed, for example, if a code generator would always minimize the reference model and then create the FSM states and transitions, but leave the addition of input/output labels to the developers.

Enterprises planning to introduce model-based testing, however, need to observe two aspects which are crucial for the success of MBT:

■ The "conventional" verification and validation workflow needed to be adapted.
■ Test engineers developing test models need higher skills than engineers programming test procedures.

The verification and validation workflow, as it is considered as state of practice today when testing according to the conventional approach, is shown in the Figure 19.10.

The requirements against which a testing campaign should be developed are used as input, potentially complemented by design documents including interface specifications. Based on these inputs, test cases are identified in a manual way and traced back to the requirements that they help to verify. These test cases are usually described in a high-level manner, so that a second step is needed to compute concrete test data and associated expected results. Then test procedures are programmed in some script language. Up to this stage, the process is manual. The following steps, however, are usually automated. The test procedures are executed against the SUT, the test verdicts are calculated, and the test results are documented and traced back to the test cases and from there to the requirements.

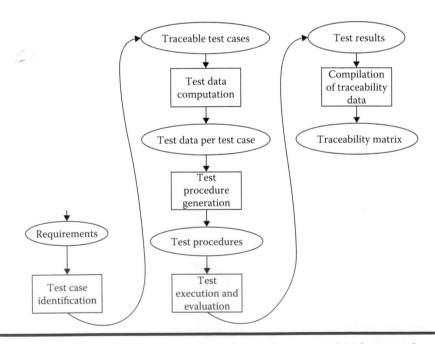

Figure 19.10 Workflow for testing according to the conventional approach.

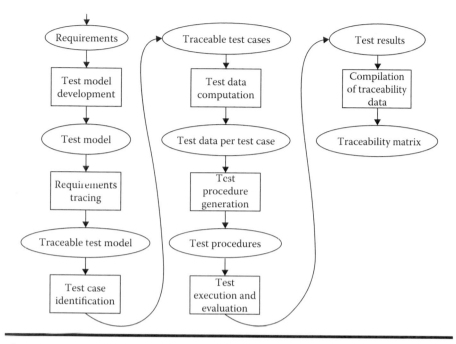

Figure 19.11   Model-based testing workflow.

The workflow for model-based testing is extended in the way shown in Figure 19.11.

When applying MBT for a testing campaign, the flow shown in Figure 19.10 is extended by two crucial steps to be performed at the beginning of the campaign: a test model has to be developed, and its elements need to be traced back to the requirements. This is a manual step, but as a return of investment, test case identification, test data computation, and test procedure generation are performed in an automated way. For safety-critical or business-critical applications, these changes in the workflow need to be well understood and documented in the quality management system of the enterprise.

Finally, it is necessary to point out that MBT engineers need the appropriate skills for developing test models. Modeling requires some abstraction capabilities that are not needed when programming test scripts. These abstractions need to target the right level: if too low, test models will turn out to be unnecessarily complex and therefore hard to validate and difficult to maintain over multiple regression-testing campaigns. If the level of abstraction is too high, important behavioral aspects of the SUT may be lost, and, consequently, the resulting test procedures will not have sufficient test strength to uncover critical implementation errors.

It should be pointed out, however, that taking care of these critical aspects results in significant returns of investments:

- Instead of needing a very large team for performing complex testing campaigns, much smaller teams of well-skilled test engineers can do an even better job, because the tool-supported MBT approach results in test suites with higher test strength.
- The test model maintenance for regression testing requires far less effort than updating huge libraries of test procedures for a new SUT release. Since all test procedures are generated from the model when applying the MBT approach, no maintenance is required on procedure level.

References

[Binder 2000]

Binder, Robert V., *Testing Object-Oriented Systems: Models, Patterns, and Tools*. Addison-Wesley, Reading, MA, 2000.

[Huang and Peleska 2016]

Huang, Wen-ling and Jan Peleska, Complete model-based equivalence class testing. *International Journal on Software Tools for Technology Transfer* 2016;18(3):265–283. DOI:10.1007/s10009-014-0356-8.

[Hubner, Huang, and Peleska 2015]

Hübner, Felix, Wen-ling Huang, and Jan Peleska, Experimental Evaluation of a Novel Equivalence Class Partition Testing Strategy. In Jasmin Christian Blanchette and Nikolai Kosmatov (eds.): *Tests and Proofs—9th International Conference, TAP 2015, Held as Part of STAF 2015*, L'Aquila, Italy, July 22–24, 2015. Proceedings. Lecture Notes in Computer Science 9154, Springer, 2015, pp. 155–172. DOI:10.1007/978-3-319-21215-9_10.

[Luo, von Bochmann, and Petrenko 1994]

Luo, Gang, Gregor von Bochmann, and Alexandre Petrenko, Test selection based on communicating nondeterministic finite state machines using a generalized Wp-method. *IEEE Transactions on Software Engineering* 1994;20(2):149–162.

Chapter 20

Open-Source Model-Based Testing Tools

One of the advantages of university teaching is that I have graduate students who are eager to try out new technologies. I used a list of model-based testing (MBT) open-source tools created by Robert V. Binder [Binder 2012] and assigned selected tools to six student teams. (Graduate students in the Grand Valley State University are typical of working professionals.) I helped each team with their investigations and attempts, but the writing (except for minor corrections) is theirs.

20.1 ModelJUnit 2.5

The ModelJUnit tool was created by Dr. Mark Utting at the University of Waikato, New Zealand. Models are written as Java classes that directly interact with a Java SUT or adapter. ModelJUnit 2.5 uses the annotations feature of Java 5.0. It then processes models to generate calls to the SUT and evaluate SUT responses (no Junit code is produced) [Binder 2012].

20.1.1 About ModelJUnit 2.5

The ModelJUnit tool is a Java Library that extends from JUnit for model-based testing and is published under the open-source license GPL v2. This had been a framework, but is now considered a library and can be used as a stand-alone tool. The tool functions by either using its own graphical user interface (GUI) or by including the library into the Java project. ModelJUnit has the ability to analyze a finite state machine written in Java and to convert it into a corresponding visual representation. After the finite state graph is created from the code, test cases can be automatically generated, customized, and executed based on user preferences.

Acquisition is simple—there is no license required in order to download ModelJUnit. There seems to be no accompanying documentation or user manual that explains the full capabilities of the ModelJUnit software. Even though the project is presented on three websites, each hosting a different version of the tool, and none contains a working version of documentation. The three web addresses are listed below:

> http://sourceforge.net/projects/modeljunit/ version 2.5, Current website of the project
> http://www.cs.waikato.ac.nz/~marku/mbt/modeljunit/ version 2.0, beta 1. Original project, hosted by university of waikato
> http://modeljunit.sourceforge.net, Current info-website of the project

The tool software seems to be abandoned, since the last update occurred on January 13, 2014, and the website "documentation" was last updated on May 15, 2009. Because of this, the bugs and suggested updates mentioned above are not likely to be fixed.

20.1.2 Using ModelJUnit 2.5 on the Garage Door Controller

Even though the implementation of the garage door opener did not seem to be too complicated, several problems occurred. The main challenge was to force the ModelJUnit tool to accept the GarageDoorOpener-FSM definition. This was finally achieved by compiling the entire ModelJUnit source code with our example, which does not seem to be the intended course of action to get into the tool. Once the tool loaded the FSM and the tests have been generated, all states appear as expected, however, in trying out different configurations; the events *Laser_Crossed* (*e5*) and *Obstacle_Hit* (*e4*) were seldom activated. Furthermore when the "show unexplored states/actions" in the label section is activated, more states and actions appear than expected and the 'missing' events *Laser_Crossed* and *Obstacle_Hit* appear. Due to the lack of documentation and understanding of the nature of this problem, the current project maintainer, Mark Utting, was contacted, and he did offer assistance.

Here is the Java code to describe the Garage Door Controller finite state machine:

```
import nz.ac.waikato.modeljunit.Action;
import nz.ac.waikato.modeljunit.FsmModel;
import nz.ac.waikato.modeljunit.RandomTester;
import nz.ac.waikato.modeljunit.Tester;
import nz.ac.waikato.modeljunit.coverage.CoverageMetric;
import nz.ac.waikato.modeljunit.coverage.TransitionCoverage;

public class GarageDoorOpener implements FsmModel
```

```
{
    private STATE state;

    public enum STATE
    {
        DOOR_UP,                    //Door is currently up
        DOOR_DOWN,                  //Door is currently down
        DOOR_STOPPED_DOWN,          //Door stopped while moving down
        DOOR_STOPPED_UP,            //Door stopped while moving up
        DOOR_CLOSING,               //Door currently closing
        DOOR_OPENING                //Door currently opening
    }

    private MOTOR_STATE motor_state;

    public enum MOTOR_STATE
    {
        MOTOR_UP,                   //Motor going up
        MOTOR_DOWN,                 //Motor going down
        MOTOR_STOPPED               //Motor stopped
    }

    public Object getState() {
        return state;
    }

    public GarageDoorOpener()
    {
        reset(true);
    }

    public void reset(boolean testing) {
        state = STATE.DOOR_DOWN;
        motor_state = MOTOR_STATE.MOTOR_STOPPED;
    }

    @Action public void Control_Signal()
    {
        if(state == STATE.DOOR_UP)
        {
            state = STATE.DOOR_CLOSING;
            motor_state = MOTOR_STATE.MOTOR_DOWN;
            return;
        }
```

```
if(state == STATE.DOOR_DOWN)
{
    state = STATE.DOOR_OPENING;
    motor_state = MOTOR_STATE.MOTOR_UP;
    return;
}

if(state == STATE.DOOR_STOPPED_DOWN)
{
    state = STATE.DOOR_CLOSING;
    motor_state = MOTOR_STATE.MOTOR_DOWN;
    return;
}

if(state == STATE.DOOR_STOPPED_UP)
{
    state = STATE.DOOR_OPENING;
    motor_state = MOTOR_STATE.MOTOR_UP;
    return;
}

if(state == STATE.DOOR_CLOSING)
{
    state = STATE.DOOR_STOPPED_DOWN;
    motor_state = MOTOR_STATE.MOTOR_STOPPED;
    return;
}

if(state == STATE.DOOR_OPENING)
{
    state = STATE.DOOR_STOPPED_UP;
    motor_state = MOTOR_STATE.MOTOR_STOPPED;
    return;
}
}

public boolean End_Of_Down_TrackGuard() {return state ==
state.DOOR_CLOSING;}
@Action public void End_Of_Down_Track()
{
    state = STATE.DOOR_DOWN;
    motor_state = MOTOR_STATE.MOTOR_DOWN;
    return;
}
```

```
public boolean End_Of_Up_TrackGuard() {return state ==
state.DOOR_OPENING;}
@Action public void End_Of_Up_Track()
{
        state = STATE.DOOR_UP;
        motor_state = MOTOR_STATE.MOTOR_STOPPED;
        return;
}

public boolean Obstacle_HitGuard() {return state == state.DOOR_CLOSING;}
@Action public void Obstacle_Hit()
{
        state = STATE.DOOR_OPENING;
        motor_state = MOTOR_STATE.MOTOR_UP;
        return;
}

public boolean Laser_CrossedGuard() {return state == state.DOOR_CLOSING;}
@Action public void Laser_Crossed()
{
        state = STATE.DOOR_OPENING;
        motor_state = MOTOR_STATE.MOTOR_UP;
        return;
}

public static void main(String[] args)
{
    Tester tester = new RandomTester(new GarageDoorOpener());
    tester.buildGraph();

    CoverageMetric trCoverage = new TransitionCoverage();
    tester.addListener(trCoverage);
    tester.addListener("verbose");
    tester.generate(20);

    tester.getModel().printMessage(trCoverage.getName() + "was"+
    trCoverage.toString());
    }
}
```

20.1.3 General Comments

ModelJUnit provides a lot of functionality in an appealing combination of a library and a GUI tool. It provides several useful examples to get started but does not explain how to get customized finite state machines into the program. The only documentation consists of the comments in the examples. This makes the process really cumbersome and leaves the user questioning the proper usage of the tool, rather than actually testing the FSM of interest. Furthermore, the application is buggy and freezes from time to time.

This section is derived from a report by Grand Valley State University graduate students Lisa Dohn, Roland Heusser, and Abinaya Muralidharan.

20.2 Spec Explorer

Spec Explorer was developed within the Microsoft organization. It is a Visual Studio add-in that generates test suites from C# model programs; it provides visualization and modeling features [Binder 2012]. The GVSU students report that it has not been supported in the .Net environment since 2012.

20.2.1 About Spec Explorer

Spec Explorer is an add-in tool for Visual Studio that features Model Based Testing (MBT) functionality in the .NET environment. MBT with Spec Explorer allows testers/developers to model their applications using C# or Visual Basic and to generate state machine diagrams and unit tests from those models. Spec Explorer can be downloaded for free from MSDN, and the installation is client-by-client. It is very simple to use—it does not require any special configuration; however, learning the tool can be difficult, and it may take some time before one understands the level of abstraction of Spec Explorer. The tool is well documented on MSDN, and developers can also find useful examples online. Spec Explorer is very powerful because it automatically generates test cases from very simple code that represents the model of the application. Currently Spec Explorer is available for Visual Studio 2010 and 2012, and Microsoft did not upgrade it for 2013 or the 2015 version Visual Studio.

20.2.2 Using Spec Explorer

Installing Spec Explorer adds the project template "Spec Explorer Model" in Visual Studio, under the subcategory "Test," in the "New Project" screen. This is the starting point for developing an model based testing project. The full-blown application of the Garage Door Controller is composed of several classes that represent the different components of the application (such as Control Device, Light Beam Sensor, etc.). A Spec Explorer Model solution is composed of three main projects:

- *Implementation*: The code of the application
- *Models*: The intended behavior to analyze
- *TestSuite*: The Unit Test project that contains the test cases automatically generated by the tool

Modeling the Garage Door Controller application required several "trials and errors" before the state machine looked exactly the way the model was intended. Spec Explorer assumes a certain level of familiarity with .NET languages (we used C#), as well as with the Visual Studio environment. This can be an obstacle when it comes to using Spec Explorer for the first time. The team spent a total of 28 hours to complete the Garage Door example.

20.2.2.1 Implementation

For the Garage Door Controller implementation, we created a simplified version of the application: the entire program is condensed into one class called GarageDoorImplementation that contains the various events (for instance, ControlDevice_ButtonPressed) and actions:

```csharp
using System;

namespace GarageDoorMBT.Implementation
{
    public class GarageDoorImplementation
    {
        private static MotorState _state;

        #region Events

        public static void ControlDevice_ButtonPressed()
        {
            switch (_state)
            {
                case MotorState.Going_Up:
                    StopMotorUp();
                    break;

                case MotorState.Going_Down:
                    StopMotorDown();
                    break;

                case MotorState.Stopped_While_Up:
                    StartMotorUp();
                    break;

                case MotorState.Stopped_While_Down:
                    StartMotorDown();
                    break;
```

```
          case MotorState.Stopped_End_Up:
              StartMotorDown();
              break;

          case MotorState.Stopped_End_Down:
              StartMotorUp();
              break;

          case MotorState.Stopped:
              StartMotorUp();
              break;
      }
}

public static void TrackEndUpSensor_EndReached()
{
    if (_state == MotorState.Going_Up)
        StopMotorEndUp();
}

public static void TrackEndDownSensor_EndReached()
{
    if (_state == MotorState.Going_Down)
        StopMotorEndDown();
}

public static void LightBeam_BeamInterrupted()
{
    if (_state == MotorState.Going_Down)
    {
        StopMotorDown();
        StartMotorUp();
    }
}

public static void ObstacleSensor_ObstacleEncountered()
{
    if (_state == MotorState.Going_Down)
    {
        StopMotorDown();
        StartMotorUp();
    }
}
```

```
#endregion Events

#region Actions

private static void StopMotorUp()
{
    _state = MotorState.Stopped_While_Up;
}

private static void StopMotorDown()
{
    _state = MotorState.Stopped_While_Down;
}

private static void StopMotorEndUp()
{
    _state = MotorState.Stopped_End_Up;
}

private static void StopMotorEndDown()
{
    _state = MotorState.Stopped_End_Down;
}

private static void StartMotorUp()
{
    _state = MotorState.Going_Up;
}

private static void StartMotorDown()
{
    _state = MotorState.Going_Down;
}

#endregion Actions

public override string ToString()
{
    string format = "[S{0}] Motor State => {1}";
    return String.Format(format, (int)_state, _state.ToString().Replace("_", " "));
}
    }
}
```

In contrast, the full-blown application of the Garage Door Controller is composed of several classes that represent the different components of the application (such as Control Device and Light Beam Sensor). For the purpose of Spec Explorer and Model Based Testing, we can omit most of these components and work only on the state behavior of the application.

The events/transitions of the model (in the implementation class) must be defined as static methods; the same methods will appear as "rules" in the Model class.

Note: The ToString() method is particularly useful in the Explorer (state machine) diagram, and it should return a string of the current status of the application.

The GarageDoorModelState class is the placeholder of the actual state of the application:

```
namespace GarageDoorMBT.Implementation
{
    public struct GarageDoorModelState
    {
        public MotorState State;
    }

    /// <summary>
    /// Enum that represents the possible states of the motor
    /// </summary>

    public enum MotorState
    {
        Going_Up = 1,
        Going_Down = 2,
        Stopped_While_Up = 3,
        Stopped_While_Down = 4,
        Stopped_End_Up = 5,
        Stopped_End_Down = 6,
        Stopped = 7, // Unknown state of the motor
    }
}
```

20.2.2.2 The Spec Explorer Model

The Model (GarageDoorModel) class contains the intended behavior of the application. It is an abstraction of the Implementation class. This class contains all the rules that will be used by Spec Explorer to generate the state machine. These rules are flagged with the [Rule] attribute, and they are presented as static methods with the same signature as the ones in the Implementation class.

```csharp
using GarageDoorMBT.Implementation;
using Microsoft.Modeling;

namespace GarageDoorMBT.Models
{
    public static class GarageDoorModel
    {
        // Model state
        public static GarageDoorModelState GarageDoor =
        new GarageDoorModelState()
        {
            State = MotorState.Stopped
        };

        [Rule]
        public static void ControlDevice_ButtonPressed()
        {
            switch (GarageDoor.State)
            {
                case MotorState.Going_Up: // S1
                    GarageDoor.State = MotorState.Stopped_While_Up; // S3
                    break;

                case MotorState.Going_Down: // S2
                    GarageDoor.State = MotorState.Stopped_While_Down; // S4
                    break;

                case MotorState.Stopped_While_Up: // S3
                    GarageDoor.State = MotorState.Going_Up; // S1
                    break;

                case MotorState.Stopped_While_Down: // S4
                    GarageDoor.State = MotorState.Going_Down; // S2
                    break;

                case MotorState.Stopped_End_Up: // S5
                    GarageDoor.State = MotorState.Going_Down; // S2
                    break;

                case MotorState.Stopped_End_Down: // S6
                    GarageDoor.State = MotorState.Going_Up; // S1
                    break;

                case MotorState.Stopped: // S7
                    GarageDoor.State = MotorState.Going_Up; // S1
                    break;
```

```
        }
    }

    [Rule]
    public static void TrackEndUpSensor_EndReached()
    {
        Condition.IsTrue(GarageDoor.State == MotorState.Going_Up);
        if (GarageDoor.State == MotorState.Going_Up)// S2
            GarageDoor.State = MotorState.Stopped_End_Up; // S5
    }

    [Rule]
    public static void TrackEndDownSensor_EndReached()
    {
        Condition.IsTrue(GarageDoor.State == MotorState.Going_Down);
        if (GarageDoor.State == MotorState.Going_Down)// S2
            GarageDoor.State = MotorState.Stopped_End_Down; // S6
    }

    [Rule]
    public static void LightBeam_BeamInterrupted()
    {
        Condition.IsTrue(GarageDoor.State == MotorState.Going_Down);
        if (GarageDoor.State == MotorState.Going_Down)// S2
            GarageDoor.State = MotorState.Going_Up; // S1
    }

    [Rule]
    public static void ObstacleSensor_ObstacleEncountered()
    {
        Condition.IsTrue(GarageDoor.State == MotorState.Going_Down);
        if (GarageDoor.State == MotorState.Going_Down)// S2
            GarageDoor.State = MotorState.Going_Up; // S1
    }
    }
}
```

20.2.2.3 The Coordination File

The coordination (.cord) file is the most important part of the Spec Explorer Model project. It contains the definitions (in the form of scripts) for the state machines and the test cases that will be automatically generated. It is composed of the Main block: here the developer specifies which rules (from the implementation class) are included (in this example all), as well as additional parameters such as the path and namespace for the generated test cases. The machine block represents the simulated behavior of the application. Machines can be used to generate state machine diagrams and/or test cases. The machines can depict the entire application (see machine GarageDoorModel) or specific transitions (see machine CustomScenario).

```
// This is a Spec Explorer coordination script (Cord version 1.0).
// Here is where you define configurations and machines describing the
// exploration to be performed.

using GarageDoorMBT.Implementation;

/// Contains actions of the model, bounds, and switches.
config Main
{
    // Use all actions (rules) from the implementation class
    action all GarageDoorMBT.Implementation.GarageDoorImplementation;

    switch StepBound = none;
    switch PathDepthBound = none;
    switch StateBound = 250;

    switch TestClassBase = "vs";
    switch GeneratedTestPath = "..\\GarageDoorMBT.TestSuite";
    switch GeneratedTestNamespace = "GarageDoorMBT.TestSuite";
    switch TestEnabled = false;
    switch ForExploration = false;
}


// Model for simulating simple operations
machine GarageDoorModel() : Main where ForExploration = true
{
    construct model program from Main
    where scope = "GarageDoorMBT.Models.GarageDoorModel"
}


machine CustomScenario() : Main where ForExploration = true
{
    //Omitting the parenthesis for an action invocation
    //is equivalent to setting all its parameters to _ (unknown).
    (ControlDevice_ButtonPressed; ControlDevice_ButtonPressed;
    ControlDevice_ButtonPressed; ControlDevice_ButtonPressed;
    ControlDevice_ButtonPressed)*
}
```

```
// Test suite
machine GarageDoorMBTTestSuite() : Main where ForExploration = true,
TestEnabled = true
{
    construct test cases
    where strategy = "ShortTests"
    for GarageDoorModel()
}
```

20.2.2.4 Tool Execution

With the Spec Explorer project completed as described above, the tester can explore and analyze the behavior of the model. The "Exploration Manager" screen shows the machines defined in the cord file:

From the "Explorer Manager," right-click on a machine that is "Test Enabled" to generate test cases (the code will be saved in the Unit Test project).

This is an example of unit test code automatically generated from the model:

```
#region Test Starting in S0
[Microsoft.VisualStudio.TestTools.UnitTesting.TestMethodAttribute()]
public void GarageDoorMBTTestSuiteS0() {
    this.Manager.BeginTest("GarageDoorMBTTestSuiteS0");
    this.Manager.Comment("reaching state \'S0\'");
    this.Manager.Comment("executing step \'call
    ControlDevice_ButtonPressed()\'");
    GarageDoorMBT.Implementation.GarageDoorImplementation.
    ControlDevice_ButtonPressed();
    this.Manager.Comment("reaching state \'S1\'");
    this.Manager.Comment("checking step \'return
    ControlDevice_ButtonPressed\'");
    this.Manager.Comment("reaching state \'S10\'");
    this.Manager.Comment("executing step \'call
    ControlDevice_ButtonPressed()\'");
    GarageDoorMBT.Implementation.GarageDoorImplementation.
    ControlDevice_ButtonPressed();
    this.Manager.Comment("reaching state \'S15\'");
    this.Manager.Comment("checking step \'return
    ControlDevice_ButtonPressed\'");
    this.Manager.Comment("reaching state \'S20\'");
    this.Manager.Comment("executing step \'call
    ControlDevice_ButtonPressed()\'");
    GarageDoorMBT.Implementation.GarageDoorImplementation.
    ControlDevice_ButtonPressed();
```

```
    this.Manager.Comment("reaching state \'S25\'");
    this.Manager.Comment("checking step \'return
    ControlDevice_ButtonPressed\'");
    this.Manager.Comment("reaching state \'S30\'");
    this.Manager.EndTest();
}
#endregion
```

The test cases can be run against the implementation class, or they can be exported into a separate Unit Test project and run against the actual application. The latter will require a minor adaption due to the potentially different namespaces and class names.

20.2.3 General Comments

Installing Spec Explorer is trivial and requires very few "clicks" in the installation wizard. The learning process takes some time, though. From a tester's point of view, the level of abstraction of the application to model may not be quite clear at the beginning. Once this is understood, creating the model with the code does not take too much time (the code itself can be very small). It also takes some time to understand the concepts of "machine" and "rule;" after that, the creation of the model and the cord scripts becomes fairly easy (proportional to the complexity of the model).

Spec Explorer is simple to use and fairly intuitive. Documentation is available online on MSDN, as well as a couple of tutorials/examples. A developer can model an application by creating the code that is just enough to represent the main behavior of the application. From a tester's perspective, though, it may take some time to understand the level of abstraction in Spec Explorer, and it may take a couple of "attempts" before seeing a state machine diagram that corresponds to the intended behavior of the model. In conclusion, Spec Explorer is a very powerful tool for Model Based Testing as it allows testers to generate test cases from a model of an application.

This section is derived from a report by Grand Valley State University graduate students Khalid Alhamdan, Frederic Paladin, Saheel Sehgal, and Mike Steimel.

20.3 MISTA

MISTA (for Model-Based Integration and System Test Automation) was developed at Boise State University in Idaho. It is available at no charge for academic use and is also available commercially. It uses lightweight high-level Petri nets as a visual modeling notation. Test models can be animated and verified. MISTA generates executable test code from a test model in Java, C, C++, C#, VB, or HTML and/or test engines JUnit, NUnit, and Selenium [Binder 2012].

20.3.1 About MISTA

Acquiring MISTA is simple: go to the developer's website, http://cs.boisestate.edu/~dxu/research/MBT.html, and click the MISTAv1.0 download link to download a zip file of the program. Then simply extract the folder from the zip file in the desired location. No licensing is specified in any of the documentation for the tool.

20.3.1.1 MISTA Environment

MISTA is a Java jar application. The documentation does not specify if there are any version requirements for Java. The tool itself provides a GUI that allows users to produce the models, a spreadsheet-like editor to textually make the models, options to enter helper code for test code generation and test tree generation. It also allows the user map portions of the model to the constructs used for test code generation. The graphical editor for model development is based on of PIPE3 (Platform Independent Petri Net Editor).

20.3.1.2 MISTA Capabilities

One useful feature of the tool is model verification. The tool will check to make sure that all states and transitions in a model can be reached, given the initial and goal state. This allows the user to have confidence that the model is correct. The tool also allows for Petri nets to be simulated to make sure the model works as expected. Generating test cases in the tool is as simple as picking the test coverage decided and then clicking a 'Generate Test Code' button. The tool provides a wide variety of test coverage options and criteria. Below is the list of criteria taken from the user manual:

- *Reachability tree coverage*: MISTA first generates the reachability graph of the function net with respect to all given initial states and, for each leaf node, creates a test from the corresponding initial state node to the leaf.
- *Reachability + invalid paths (sneak paths)*: MISTA generates an extended reachability graph—for each node, MISTA also creates child nodes of invalid firings (they are leaf nodes). A test from the corresponding initial marking to such a leaf node is called a dirty test.
- *Transition coverage*: MISTA generates tests to cover each transition.
- *State coverage*: MISTA generates tests to cover each state that is reachable from any given initial state. The test suite is usually smaller than that of reachability tree coverage because duplicate states are avoided.
- *Depth coverage*: MISTA generates all tests whose lengths are no greater than the given depth.
- *Random generation*: MISTA generates tests in a random fashion. The parameters used as the termination condition are the maximum depth of tests and the maximum number of tests. When this menu item is selected, the user will be asked to set up the maximum number of tests to be generated. The actual

number of tests is not necessarily equal to the maximum number because random tests can be duplicated.

- *Goal coverage*: MISTA generates a test for each given goal that is reachable from the given initial states. Before generating tests for this coverage, you should verify goal reachability to see if they are reachable. Typically, the firing sequences that reach the given goals will be transformed into tests.
- *Assertion counterexamples*: MISTA generates tests from the counterexamples of assertions that result from assertion verification. Before generating tests, the user may verify assertions to see if the specified assertions have counterexamples.
- *Deadlock/termination states*: MISTA generates tests that reach each deadlock/ termination state in the function net. A deadlock/termination state is a marking under which no transition can be fired. Test generation makes use of the result of "Check for Deadlock/Termination States".
- *Given sequences*: MISTA generates tests from the firing sequences in a given file (e.g., the log file of simulation or online testing under the same folder as the MID file). The file is specified by a "SEQUENCES" annotation. Make sure all the firing sequences in the file were created from the same version of the function net; otherwise test generation may fail.

20.3.1.3 Learning to Use MISTA

The learning curve for MISTA depends on two things. First, the user's familiarity with Petri nets and/or finite state machines. Second, becoming familiar with the MISTA environment itself. Not having knowledge of the first requirement raises the curve significantly. Though well developed, the MISTA environment itself can be somewhat challenging. The tools to make the models, while relatively simple, can be somewhat cumbersome. For example to rotate a transition, after adding it, you must right-click the transition, click edit, then select the relative amount you want to rotate by from a drop-down menu, and then apply it. Fortunately it comes with two documents, MISTA in a nutshell and the MISTA user's manual, which are very helpful. There are sections on varying types of Petri nets that the user can utilize in the environment. The user manual is only 56 pages, a good indication of the relative simplicity of the tool, especially when compared to commercial model-based testing tools. One plus is that the tool provides tutorial type projects that can be opened to assist with learning about Petri nets about the tool.

20.3.2 Using MISTA

20.3.2.1 The Garage Door Controller

MISTA is best suited to the Garage Door Opener problem—it is overkill for the Insurance Premium Problem. We chose to make both a finite state machine and a Petri net version of the example to explore the functionality of the tool. For the finite state machine, it took about one hour to get the model to a satisfactory state. The Petri net model took approximately 1.5 hours. Since our example is a basic problem definition without any code or program design, implementation was a little shorter, as adding helper code and MIM (model implementation mapping) would add to the time required to make the model.

20.3.2.2 Test Code Generated by MISTA

```
import junit.framework.*;

public class GarageDoorTester_RT extends TestCase{

    private GarageDoor garagedoor;

    protected void setUp() throws Exception {
        garagedoor = new GarageDoor();
    }

    public void test1() throws Exception {
        System.out.println("Test case 1");
        garagedoor.device_signal(); //constraint: Open
        assertTrue("1_1", garagedoor.Closing());
        Closing();
        garagedoor.device_signal(); //constraint: Closing
        assertTrue("1_1_1", garagedoor. Stopped_motor_engaged_down());
        Stopped_motor_engaged_down();
        garagedoor.device_signal(); //constraint:
        Stopped_motor_engaged_down
        assertTrue("1_1_1_1", garagedoor.Closing());
        Closing();
    }

    public void test2() throws Exception {
        System.out.println("Test case 2");
        garagedoor.device_signal(); //constraint: Open
        garagedoor.light_beam_interuption(); //constraint: Closing
        assertTrue("1_1_2", garagedoor. Opening());
        Opening();
        garagedoor.device_signal(); //constraint: Opening
        assertTrue("1_1_2_1", garagedoor. Stopped_motor_engaged_up());
        Stopped_motor_engaged_up();
        garagedoor.device_signal(); //constraint: Stopped_motor_engaged_up
        assertTrue("1_1_2_1_1", garagedoor. Opening());
        Opening();
```

```
    }

    public void test3() throws Exception {
        System.out.println("Test case 3");
        garagedoor.device_signal(); //constraint: Open
        garagedoor.light_beam_interuption(); //constraint: Closing
        garagedoor.end_of_up_track_reached(); //constraint: Opening
        assertTrue("1_1_2_2", garagedoor. Open());
        Open();
    }

    public void test4() throws Exception {
        System.out.println("Test case 4");
        garagedoor.device_signal(); //constraint: Open
        garagedoor.obstacle_sensor_tripped(); //constraint: Closing
        assertTrue("1_1_3", garagedoor. Opening());
        Opening();
    }

    public void test5() throws Exception {
        System.out.println("Test case 5");
        garagedoor.device_signal(); //constraint: Open
        garagedoor.end_of_down_track_reached(); //constraint: Closing
        assertTrue("1_1_4", garagedoor. Closed());
        Closed();
        garagedoor.device_signal(); //constraint: Closed
        assertTrue("1_1_4_1", garagedoor. Opening());
        Opening();
    }
}
```

20.3.2.3 Test Output Generated by MISTA

Table 20.1 summarizes the output from the various criteria for the garage door opener.

20.3.3 General Comments

We would recommend the use of the tools for model based testing. It is relatively simple, especially when compared to some commercial tools. It used standard finite state machines and Petri nets, rather than a custom modeling language, meaning it is more accessible. Petri nets are also powerful modeling tools, meaning that MISTA can be used to model a wide variety of systems. Also, the ability to automatically create test cases from the model makes the tool well suited for test-driven development. The total time spent by the team was about 20 hours.

Table 20.1 Generated Test Cases for Stopping Criteria

Test Coverage Criterion	Finite State Machine	Petri Net
Reachability tree	5	8
Reachability + invalid paths	35	134
Transition	5	8
State	4	8
Depth (=10)	328	8
Random	User defines	7 (20 requested)
Goal	(Did not work)	(Multiple)
Assertion	n/a	Not used
Deadlock	n/a	4
Given sequence	n/a	Not used

This section is derived from a report by Grand Valley State University graduate students—James Cornett, Ryan Huebner, Evgeny Ryzhkov, and Chris Taylor.

20.4 Auto Focus 3

Auto Focus 3 supports modeling and analyzing the structure and behavior of distributed, reactive, and timed computer-based systems [Binder 2012]. Apparently, it was developed in Germany, but a Google search reveals only a few academic papers.

20.4.1 About Auto Focus 3

The developers of Autofocus 3 (AF3) program describe it as a "model based development tool for distributed, reactive, embedded software systems." They also state "AF3 uses models in all development phases including requirements analysis, design of the logical architecture, and design of the hardware architecture, implementation, and deployment." AF3 is described as a plug-in for the Eclipse JDE; acquiring the software was easy-just a simple download from the developer's website, af3.fortiss.org/download/. AF3 software is supported on Windows, OS X, as well as Linux for both 32 and 64 bit machines. The AF3 software turns out not to be a plug-in for Eclipse but rather a piece of software based on the Eclipse JDE.

The list below contains the main features that are stated on the developer's website, http://af3.fortiss.org/main-features/

1. Requirement specification and analysis
   - Model-based integrated requirements analysis (MIRA)
   - Glossary and requirements source
   - Requirements specification
   - Requirement hierarchy
   - Integration with architecture
   - Requirements analysis and verification
   - Reports
2. Modeling and simulation
   - Simulate test cases
   - Modeling architecture
   - Express behavior using state automata, source code, or tables
   - Linking behavior elements to requirements
   - Animation of simulations
3. Code generation and deployment
   - Generate C code
   - New deployment
   - Code generation for deployments
4. Formal verification
   - Model checking with verification patterns
   - Counterexamples provided when verification do not pass
   - Black-box specification
   - MSC feasibility checks
   - Check for nondeterminism
   - Reachability analysis
   - Model proximity
5. Design space exploration
   - Scheduling synthesis
   - Deployment synthesis
6. Testing
   - Specify test strategy
   - Generate test suites
   - Simulate test cases
   - Easily update test suite with model changes
   - Coverage report
   - Refine test cases from requirement level to native code level
7. MBT framework that includes:
   - Generating test suites from models from given coverage criteria and input profiles
   - Simulate the model using test cases
   - Updating the test suite when the model is changed

8. Models support
  - Finite state machines
  - Mode Automata
  - State/mode transitions tables
  - Function tables

20.4.2 Using Auto Focus 3

Along with the numerous examples provided in the website, there is a tutorial on how to use the software, which gave the team high hopes for the ability to modify the examples over to the Insurance Premium Problem and to the Garage Door controller. Unfortunately, the tutorial on how to use the software was less than educational. It only gives a high-level overview of how to use the software. For example, the tutorial would often tell the user to create something, but the tutorial did not actually explain how to do this.

20.4.3 General Comments

Potential AF3 users will spend a considerable amount of time on self-education trying to learn on how to use the software tool. The student team spent 100 hours trying to use the AF3 system, with little success at testing either the Garage Door Controller or the Insurance Premium Problem. The level of detail that needs to go into the building of the model is both tremendous and confusing.

This section is derived from a report by Grand Valley State University graduate students Khalid Almoqhim, Jacob Pataniczak, and Komal Sorathiya.

20.5 Graphwalker

Graphwalker was developed at the Massachusetts Institute of Technology (MIT); it is maintained by the GraphWalker 3 Group. It generates offline and online test sequences from state machines modeled with GraphML—no UML to learn. The user selects any of the seven built-in coverage (stopping) criteria for test generation. The tool may be integrated with a Java test harness or called with SOAP as web service from the developer's server. Use of graphics to represent models limits scalability [Binder 2012].

20.5.1 About Graphwalker

Graphwalker is an open-source tool being offered under the MIT license. Tests are generated based on a model stored in GraphML form. These graphs can be generated using yEd (www.yworks.com/products/yed). It is designed to integrate with JAVA and Maven and can be used to generate tests that can be run using a test tool like JUnit or Selenium (http://graphwalker.github.io/features/). The end result is that, in theory, it can serve as a critical component in a system, which will generate and run a complete set of tests on the designed software.

Many open-source tools suffer from relatively weak documentation and complex installations requiring in-depth knowledge about each of the various obscure tools being used. Graphwalker is definitely not an exception. Although a stand-alone jar file is available, instructions on using this are not clear, and Graphwalker requires a significant amount of knowledge to actually use it. The 'how to' page used to get started offers a minimal set of instructions. There is only a small support community for Graphwalker (there is not even a Graphwalker tag in StackOverflow).

20.5.2 Using Graphwalker

In order to use Graphwalker, the user must have Apache Maven and several additional plugins. This installation requires the would-be user to wander over much of the web, to make several modifications to windows environmental variables, and to modify the security settings just to run the examples given. These examples are not too helpful either given that they were tests of some websites. Running the stand-alone jar file by itself was not helpful either as it tended to offer either minimal responses or just refer the user to the help file.

Despite the difficulty in setting up the environment, the experience of setting up the other critical tool, yEd, was quite painless. Installation was simple and painless, and the tool was very simple to use. It is a fantastic lightweight tool for designing flowcharts, and it automatically saved them in the compatible GraphML format, so there was no need to do any sort of export. The graph design using yEd turned out to be a rather simple process. Initially, the flowchart version of the Insurance program was recreated in yEd. The tool performed well and was a highly effective means of generating this flowchart. There was relatively little frustration, and the end result was pretty reasonable. Note that it also is very simple to copy the graph directly onto the clipboard and use it in the document. This was a useful exercise, but the actual requirement for input into Graphwalker was a finite state machine, which followed very specific naming conventions. Using the techniques learned when building the flowchart, it was quite simple to create the two state machines needed. It is important that the proper naming rules are used in the finite state machines. They cannot start with a number, contain spaces, or any sort of operations. Failure to do this will cause issues when trying to run Graphwalker on an "improper" model.

The test generation is done using command-line instructions. The basic instruction format contains the following steps:

■ java-jar graphwalker-cli-3.4.0.jar : This part is just executing graphwalker.
■ offline : this tells it that we are not testing a running website/program.

- ■ -- model InsuranceModelPetri.graphml : passing in the graph
- ■ "random(edge_coverage(100))" : Generator function and stop condition. This is arguably the most critical part. The choices for generator functions are somewhat limited, currently random(), quick_random(), and a_star() are the only options. The stop conditions are edge_coverage(percentage), vertex_coverage, reached_vertex, reached_edge, time_duration, never. See: http://graphwalker.github.io/generators_and_stop_conditions/

With the resulting instruction of: C:\Users\kylep\Documents\Homework>java-jar graphwalker-cli-3.4.0.jar offline--model InsuranceModelPetri.graphml "random(edge_coverage(100))"--start-element s1_Idle.

Each run of Graphwalker generated a random run through the program that followed a path until all of the edges were reached, or it could not proceed any further. This means that to thoroughly test any given state machine, it is necessary to add an additional edge from the endpoint to the startpoint.

Here is the beginning part of the generated test sequence on the insurance model (the full model output is 123 lines):

```
{"currentElementName":"s1_Idle"}
{"currentElementName":"e4_a3"}
{"currentElementName":"s2_Age_Multiplier"}
{"currentElementName":"e6_a5"}
{"currentElementName":"s3_Apply_Claims_Penalty"}
{"currentElementName":"e9_a7"}
{"currentElementName":"s4_Good_Student"}
{"currentElementName":"e11_a7"}
{"currentElementName":"s5_NonDrinker"}
{}
{"currentElementName":"s6_Done"}
```

20.5.3 General Comments

Graphwalker is a potentially very powerful tool. It offers a reasonably quick automatic test tool, and it works quite well in the environment for which it was designed, namely a Java program being managed by Apache Maven. In addition, yEd is an excellent tool. It was simple to use and took very little time to create, edit, or export graphs to pretty much anything, and it worked well with the Windows environment. The lack of documentation is problematic, and even more worrying, that there is not currently a very large support community for this, which means that there will be a lot of struggling to get GraphWalker to be operational. Given the difficulty involved in setting up the environment, it would also not necessarily be recommended to try switching to it just for the automated test capabilities provided by Graphwalker.

This section is derived from a report by Grand Valley State University graduate students Kyle Prins and Sekhar Cherukuri.

20.6 fMBT

The fMBT (free model-based testing) tool was developed by the Intel Corp. It is suitable for testing anything from individual C++ classes to GUI applications, mobile devices, and distributed systems as well as different platforms. fMBT provides a model editor, test generator, and adapters for testing through various interfaces and tools for analyzing logs [Binder 2012].

20.6.1 About fMBT

fMBT is Linux-based software, so configuration of a virtual machine is necessary. Ubuntu is a popular Linux distribution and is the recommended choice for this process. Intel provides very useful command-line snippets for installing all packages necessary for operating fMBT, but the snippets were in separate places and not easy to find: one is located in the main README file and the other on their website. The installation is rather quick, and the software must be launched from the terminal with the command 'fmbt-editor,' and the fMBT GUI would launch to allow the user to create new tests. From there, the user can create a new AAL/Python model. AAL (Adapter Action Language) is used to create tests. AAL modeling language defines models using preconditions (guard) and postconditions (body).

20.6.2 Using fMBT

The installation compressed folder came with a few examples, all of which contained models importing a class file, but it was found to be impossible as each line of Python was matched with a syntax error within the fMBT editor. To use fMBT, the user must create a model in a python class file.

If a user places the file adjacent to the running AAL file as with the examples and uses the same file import language used in those models, the user will be greeted by syntax errors for almost every line. After two hours of troubleshooting, the student team concluded that this task is more complicated than anticipated and difficult without formal instruction. The team then decided to enter everything manually, however, the tool's Python syntax is different from common Python compiler syntax. For example, the use of the "+=" operator is perfectly valid in most OOP languages but not within fMBT's editor.

The documentation available for the tool is very sparse, and the only provided tutorial was difficult to relate to the Insurance Problem. The team could also not find a way to allow for varied input, so it had to assign default, valid values to each variable. The Insurance program is quite small, and the team had tried to implement all the logic into one test step (which passed), but there was no output from the program by way of program graphs or any other advertised visualizations.

Using fMBT on the Garage Door controller was even more difficult, hence less successful.

20.6.3 General Comments

The tool was very difficult to use, especially considering users who lack Python experience. However with repeated uses with different problems, the team could see the merit in fMBT being used not only because it is open source, but also because it is very fast in developing test cases.

fMBT can be a very powerful tool and great option provided the users are familiar with the OS environments and languages. The tool offers great features for being open source, such as generation of program graphs with variable tracing and bar graphs for tracking the number of steps through the program for each block. The features and ease-of-use of fMBT are lacking compared to commercial model based testing software, but there are better free options by way of IDE plugins for Visual Studio and Eclipse. fMBT is not meant for beginner-level testers, and the documentation needs improvement. The tool is very difficult to use, especially if a user has a lack of Python experience.

This section is derived from a report by Grand Valley State University graduate students Mohamed Azuz and Ron Foreman.

Reference

[Binder 2012]
Binder, Robert V., blog, http://robertvbinder.com/open-source-tools-for-model-based-testing/, April 17, 2012.

Index

Note: Page numbers followed by f and t refer to figures and tables, respectively.